The Future of Memory

THE HISTORY OF MEDIA AND COMMUNICATION

Series Editors: Jane Rhodes and Lisa Parks

The History of Media and Communication seeks books that emphasize technology and infrastructure, the politics of social differences and intersectionality, environmental issues, globalization, and labor as they relate to media and communication. Our ambition is to publish monographs and collections that demonstrate innovative historiographic practices and engage with underexplored archives, sources, sites, voices, and experiences. We encourage attention to transnational, postcolonial, and decolonial studies through an historical lens. The series publishes research by scholars in media and communication studies whose work also engages with other fields such as gender and ethnic studies, science and technology studies, environmental studies, and global studies. The goal is to develop original, rigorous, and path-breaking books that bring forth the complex pasts of media and communication and that link to contemporary life and near futures.

For a list of books in the series, please see
our website at www.press.uillinois.edu.

The Future of Memory

A History of Lossless Format Standards in the Moving Image Archive

JIMI JONES and MAREK JANCOVIC

UNIVERSITY OF ILLINOIS PRESS
Urbana, Chicago, and Springfield

1 2 3 4 5 C P 5 4 3 2 1
∞ This book is printed on acid-free paper.

Cataloging data available from the Library of Congress

LCCN 2024060637
ISBN 978-0-252-04665-0 (cloth : alk.)
ISBN 978-0-252-08875-9 (paper : alk.)
ISBN 978-0-252-04796-1 (ebook)

Jimi dedicates this book to Karin and Jerry.

Marek dedicates this book to his parents,
Jarmila and Ján.

Contents

Acknowledgments

Jimi would like to thank all the interviewees who took the time to talk with us about their work. The work of standards development, like the work of cultural heritage preservation, is often as invisible as it is important. I hope this book helps you feel seen and appreciated for your labors. I should give a special thanks to three interviewees. The first is Dave Rice for planting the seed for this research over a decade ago when you asked, in a FADGI meeting, "Why isn't the LOC using Matroska instead of MXF?" The second is Ashley Blewer for being an ongoing, vocal support for this book. The third is James Snyder, for taking so much time to talk with me in Culpeper.

I must, of course, thank my coauthor Marek Jancovic. I have enjoyed working on this book with you and have continually been impressed by your intellect and your ability to turn my plodding prose into something engaging and at times beautiful. There is no chance that I could have done this without you. I am very proud of our collaboration.

I also extend my thanks to the people who helped me conceive of and steward this work when it was a dissertation. First, I must thank Jerome McDonough, my PhD advisor, research director, and friend. I miss you very much and have thought of you often throughout my work on this book. A man is lucky indeed if he can find father figures in his life, and you were one for me. (And I am completely certain you would have rolled your eyes at me if I had told you that.) I wish that you could see this book, but I like to think that there is some of you in it.

I would also like to thank the other members of my dissertation committee: Lori Kendall, Peter Darch, and Howard Besser. Your input into this work when it started as a dissertation was indispensable. I also thank Jody

Waitzman and Melissa Huang for their invaluable help in transcoding my interviews.

I also must give thanks and appreciation to the School of Information Sciences at the University of Illinois, Urbana-Champaign (which will, for me, remain the Graduate School of Library and Information Science) for supporting the research that is the core of this book. I would also like to thank Linda Smith for her indefatigable aid in finding funding for my research-related travel and for taking a chance on me as an instructor for the School a decade and a half ago. I also want to thank the many students I have had the pleasure to teach. You are too numerous for me to list here without this section being longer than the credits of a Marvel film, but you have all helped to give my life meaning and purpose. I hope some of you discover this book and find some enjoyment in it.

I have been fortunate to have a life full of professional and academic mentors. Mark Somers and Holli Hanley, I thank you both for giving me my start in film production. Kevin Hanson, I thank you for fostering in me both a love of filmmaking and the understanding that the most important piece of filmmaking equipment is the box of donuts for the crew. Kevin Hamilton, thank you for being a friend and for informing the design of this work in its earliest stages. Thank you, James Hay, for your ongoing support for me and my work.

I give a wealth of gratitude to Madelyn Garrett, Greg Thompson, Victoria Hindley, Luise Poulton, Roy Webb, and Jennifer Teper for your mentorship as I started my career in libraries and archives. Tawnya Keller and Snowden Becker, you get special mention because you are mentors and because you are like sisters to me. I love you both dearly. Jack Brighton, I am thankful for your support as a colleague and a friend. Thank you, Kate McDowell and Ben Grosser, for the conversations that have helped to inform this work. Bill Brand, you have my thanks for collaborating with me on the most innovative and compelling class I have ever had the pleasure to teach. I would also like to thank Martha Anderson, Beth Dulabahn, Carl Fleischhauer, Genita Coger, Trevor Owens, Kate Murray, and my other NDIIPP and FADGI colleagues for their support during my time at the Library of Congress. (Kate Murray, you get an extra special shout-out for finding Karin and me a place to live in the DC area and for helping us feel so welcome during our time there.)

Finally, I thank my family and friends for their support and their faith in me. Thank you to my parents, Karen and Harry Carson, for your belief in me through the years. Thank you to Piotr Adamczyk for your invaluable input into my research and your friendship. Thank you to my dear friend Arsheeya Mashaw for calling me "Dr. Jones" for all these decades. And I give thanks to other loving friends who have supported me through the years: Cricket, Ryan and Jasper Krengel, Tracy Medley, Angie Midgley, Ryan Edge, Susan Becker, Simon and Cameron Hamilton, Erin Miller and Jadon Peck, Evan Alvarez, Jes Park, Matt Crowther, Joe Agati, Jennifer Allen, Deke Weaver, Kim Kwee, Mark Barnes, Olive Barnes, Shane Beers, Linda Bial, Robert Dignan, Liam Moran, Robyn Bianconi, Beth Bloch, Jess Adamczyk, Laurie Chipps, Jeffrey T. Jones, Brandon Christensen, Robert Olshin, Molly Dolan, Andy Ducett, Elizabeth Edwards, Adam Fotos, Brian Fritts, Ed Glaser, Jason Gonzalez, Jenny Davis, Leslie Johnston, Joe and Sylvia Jones, Miriam Kienle, Kathryn LaBarre, Carol Tilley, Dan Larson, Ruta Abolins, Tracy Popp, Will Schlaack, Amanda McClellan, Lars Medley, Greg Midgley, Anna Oates, Rashelle Roos, Sarah Stauderman, Paul Stout, Ann Wilkens, Penny Wilson, Rulon Wood, Preston Wright, and Dan Wright.

And, most importantly, I give thanks to my spouse, best friend, and collaborator, Karin Hodgin Jones. You are, hands-down, the most brilliant person I have had the pleasure of knowing. Your optimism and love of life are an inspiration to me. Throughout this work you have shown me when to say yes and, at least as importantly, when to say no. Every time I discover something new, you are the person I want to rush home and share it with. I am so thankful to share this journey with you. "Us."

Marek wishes to thank his colleagues at the Johannes Gutenberg University of Mainz, where this project was started, and colleagues at the Vrije Universiteit Amsterdam, where it was completed. First and foremost, I thank Alexandra Schneider. Some of the initial seedlings for this book began many years ago as a chapter in my doctoral dissertation, and it was Alexandra's stellar mentorship and academically generous guidance that helped it grow into a book and a collaborative project with Jimi. Wanda Strauven's acute comments have also been instrumental and helped me spell out many of the points I was struggling to articulate. Equally important has been the input of Marc Siegel, Markus Stauff, Ginette Verstraete, and colleagues at the VU's CLUE+ Interfaculty Institute for Culture, Cognition, History, and Heritage. Sebastian Scholz, Ivo Blom, and Klaas de Zwaan—thank you for making me feel at home at the Vrije Universiteit Amsterdam under the impossible conditions of a global pandemic.

The research we have conducted for this book would have been unthinkable without the generous help of scholars and archiving and preservation experts who have taken the time to provide us with invaluable insights into their work. Besides all of the people interviewed in this book, we relied on the extensive networks and recommendations of Vinzenz Hediger,

Johan Oomen, Ilse Assmann, Benjamin Pius, and Rebecca Ohene-Asah. Giovanna Fossati, Eef Masson, and the Moving Images: Preservation, Curation, Exhibition group at the Amsterdam School for Cultural Analysis have offered us the platform where this project was really born: It was at one of the events of the research group at the EYE Film Museum Netherlands where Jimi and I met and first began talking about maybe, possibly, perhaps wanting to combine our then ongoing individual research projects on archival format standardization and the history of video compression.

Conversations with Axel Volmar and Maral Mohsenin have provided us with exciting impulses, much food for thought, and wonderful real-life examples, and the conferences of the NECS, IASA, and NTTW communities were fantastic opportunities to discuss this work with a larger audience. It was a blessing to be able to have exchanges with scholars like Nicole Braida, Elisa Linseisen, Oliver Fahle, Rosa Menkman and—once more, Alexandra and Axel—to collaboratively work through questions of media formats and formatting, which remain infinitely fascinating to me precisely because of those conversations. I also owe a good portion of gratitude to the German Research Foundation graduate research training program Configurations of Film, of which I was an associate member during part of the research process. The discussions facilitated by the program and the always brilliant, incisive feedback from its members have decisively shaped my thinking about the subject. Three anonymous reviewers have read our manuscript in its early stages, and I wish to compliment and thank them for their meticulous engagement with our work, and for the constructive and deeply valuable recommendations.

Jimi, it was such a pleasure to work on this book with you. I wouldn't exactly describe the process as easy, but having you as a coauthor made this as easy and effortless as writing an academic book can be. Thank you for your patience, your focus, and for bearing with me while I swerved and got sidetracked, sometimes for years, by unworkable ideas and argumentative dead-ends.

For their support throughout the long process of writing, I thank my dear parents, Jarmila and Ján, and my family and friends. Roy Marcks, Marije Bartholomeus, Takuma Fujii, Puya Sabet, Xiaodan Sommer-Zhang, Yuuka Yamaoka, Esteban Ramírez Hincapié, Martijn van Veen—your friendship got me through the madness of living and working in two different countries. I'm glad to have left the border-crossing commute behind me, but all the happier that you were there to keep me sane. Lastly, thank you, Micha—mon ciel, mon feu, ma flamme.

List of Acronyms

AAF	Advanced Authoring Format
AMIA	Association of Moving Image Archivists
AMWA	Advanced Media Workflow Association
AS-07	The MXF Archive and Preservation Format Application Specification, developed by the Library of Congress with AMWA
AV1	AOMedia Video 1, an open and royalty-free video compression standard
AVC	Advanced Video Coding, a video compression standard
AVI	Audio Video Interleave format, a common audiovisual file container
BBC	British Broadcasting Corporation
CELLAR	Codec Encoding for LossLess Archiving and Realtime transmission, a standardization working group within the IETF
DAT	Digital Audio Tape
DC28	The Technology Committee on Digital Cinema, a working group within SMPTE
DCI	Digital Cinema Initiative, a consortium of Hollywood studios
DPX	Digital Picture Exchange format
EBML	Extensible Binary Meta Language
EBU	European Broadcasting Union
FADGI	Federal Agencies Digital Guidelines Initiative, a collaboration of US federal agencies that develops preservation guidelines

FFV1	A lossless digital video compression and encoding method
FIAF	International Federation of Film Archives
FLAC	Free Lossless Audio Codec
HEIC	High Efficiency Image Codec, a container format for digital still images, developed by the Moving Picture Experts Group
HEVC	High Efficiency Video Coding, a common video compression standard
IASA	International Association of Sound and Audiovisual Archives
IEC	International Electrotechnical Commission
IETF	Internet Engineering Task Force
ISO	International Organization for Standardization
ITU	International Telecommunication Union
J2K	Colloquial term for JPEG 2000
JPEG 2000	A lossless digital image compression standard
LoC	The United States Library of Congress
LTO	Linear Tape Open, a common high-density tape-based data storage format
MDPI	Indiana University's Media Digitization and Preservation Initiative
MKV	The Matroska file format and its file extension
MP3	MPEG Audio Layer III, a common digital audio encoding and file container
MP4	Moving Pictures Expert Group 4, a video encoding and file container
MPEG	Moving Picture Experts Group, a working group of the International Organization for Standardization
MXF	Material Exchange Format, a container for digital multimedia information
NHK	Japan Broadcasting Corporation
PREFORMA	Preservation Formats for Culture Information/E-archives, a European project for standardizing file formats for long-term digital preservation
SAMMA	System for Automated Migration of Media Assets, a robot for digitizing video cassettes to digital files
SCOT	Social Construction of Technology
SI	Social Informatics
SMPTE	Society of Motion Picture and Television Engineers
STS	Science and Technology Studies

TC-06 Guidelines for the Preservation of Video Recordings, an IASA technical publication

TIFF Tagged Image File Format, a common container for digital still image data

VVC Versatile Video Coding, a recent video compression standard

The Future of Memory

INTRODUCTION

So Many Standards, So Little Time

The People and Politics Behind Archival Video Formats

In 2009, Hermann Lewetz, a media preservation specialist at the Austrian national sound and video archive, embarked upon a project to digitize one thousand analog video recordings to store them online.[1] Lewetz experimented with digitization guidelines created by the US Library of Congress that utilized two media technologies—the digital file container MXF (the Media Exchange Format) and the video compression standard JPEG 2000. But he could not make the test video files provided by the library play back properly. Lewetz knew he wanted to create files with the highest image fidelity to the analog originals, so he worked with fellow Austrian preservation specialist Peter Bubestinger-Steindl to cast about for an alternative to JPEG 2000 that would generate similar, if not better, files.

Buried in the free and open-source software tool FFmpeg, which is widely used to transcode or recompress video from one format to another, Bubestinger-Steindl discovered an obscure, experimental video compression algorithm called FFV1 and began running tests on it. The larger European preservation community was unaware of FFV1. This is not surprising given that, unlike JPEG 2000, it was an unstandardized algorithm of dubious origin without the backing of any recognized industry association or professional body.

But upon seeing Bubestinger's successful tests with FFV1, Lewetz soon found that it served his needs perfectly, so he began using it for his preservation work.[2] Lewetz was, in fact, so impressed with FFV1 that he decided to present about its use at the 2010 joint conference of the Association of Moving Image Archivists and International Association of Sound and Audiovisual Archives—two venerable professional organizations that unite individuals and institutions involved in media preservation globally. It was at this conference that FFV1 was unveiled to the international media preservation world as a viable preservation technology for high-quality video files. But the audience's response was not what Lewetz expected. In a later interview, he said that he experienced threats because of this presentation—in what could easily be a scene from a gangster film, a vendor physically intimidated him by harshly putting his hand on Lewetz's shoulder and ominously intoning that his company "can't allow FFV1 for its customers." In Lewetz's recounting, he was bombarded with hostile questions from the conference participants who doubted FFV1's reliability, and with appalled reactions to his choice of a video encoder that had not been ratified by formal bodies like the International Organization for Standardization (ISO).[3] The casual observer of these events might ask—perfectly logically—*why* would these adult professionals express such fear and anger over *video standards*, of all things?

Much is at stake in the work of standardizing video. The reason why you can record a video with your phone, send it through a messaging app to someone who will watch it on their computer, or upload it to a streaming platform where it can be seen by millions of people on their TVs and other devices, is because those devices speak the common language of metadata, compression algorithms, and standardized file formats. A Blu-ray Disc from Bulgaria will work on a Korean Blu-ray player in a home in Brisbane. A film producer can rest assured that the hard drive with her precious new indie gem will play without issues at a film festival half a world away—thanks to video file format and encoding standards. Standards like these are an important part of what allows local media companies to participate in global media industries. Much of our culture and entertainment depends on the hidden workings of standardized technology that operates according to well-defined rules. This is a book about how some of those rules are made and who gets to make them. At the center of the story is the complex and fascinating standardization process of the video compression and container formats JPEG 2000, FFV1, MXF, and Matroska.

If you have used a computer, you may be familiar with video formats like AVI or MP4. Few people, however, would recognize many of the acronyms used in audiovisual archives. And yet throughout the past decade, these four video standards have been at the center of a heated format war whose outcome could decide the future of our audiovisual memory. When you see a film in a cinema these days, you are experiencing the outputs of JPEG 2000 and MXF standardization: JPEG 2000 is a compression algorithm that makes image files smaller and MXF is a "wrapper" or "container" format that packages the hundreds of thousands of still frames that a film consists of into a single digital file together with sound, subtitles, and other information. Outside of the cinema, you may also have encountered pirated digital films circulating online, and chances are that most of them had the file extension MKV. This acronym designates the Matroska format, another container that many archives across the globe are beginning to use to preserve our audiovisual cultural heritage, often in combination with the compression algorithm FFV1. Standardized file formats are essential to the media industry and to contemporary moving image archives. Format standards ensure that institutions can exchange information, verify the objects in their custody, and ultimately preserve them according to internationally accepted workflows.

Traditionally, much of the development of moving image standards has been driven by Hollywood and the American and British broadcast industries. The United States Library of Congress has chosen JPEG 2000 and MXF, two industry-influenced standards, for the digitization of their video materials. But other institutions are choosing to work with the—initially controversial—open-source standards FFV1 and Matroska, which previously had no formal backing from any standards organization. The differences between these approaches offer rich grounds for exploration and inquiry because the strategic choice an archive makes when it chooses one format over another is based not only on technological or financial considerations. It may also be informed by ethical and even political reasons: Format standards revolve not just around the technicalities of video itself but also serve as proxies for notions like freedom and openness. Understanding this can show that archives are deeply social institutions. But observing the process of video format development and adoption in the archive can also teach us about the much wider, fundamental role of technological standards in contemporary society and media culture and lets us investigate the reconfigurations and shifts in power taking place within the media landscape. Studying format standards in audiovisual archives also crucially involves fields like software development or algorithms and forensic methods, whose significance in media preservation (and media theory) is yet to be fully understood.

Drawing on interviews with archivists, standards developers, and preservation experts, this book gives insight into the social processes and material difficulties involved in producing, choosing, and implementing archival format standards—the “raw data” of our visual history. This is both a history of an important decade in audiovisual preservation, as well as a snapshot of a field undergoing a tremendous transition. The powerful film and broadcast industries have traditionally dictated format standards, often with interests opposing those of archivists. The standardization process that we follow in this book, however, is unprecedented: It is the first-time archivists have begun to proactively develop their own digital video standards. This is a complex undertaking that occasionally pits smaller memory institutions with limited finances against larger and comparatively better-resourced national archives. Other new players have also entered the field: tech companies, hardware manufacturers and software developers, film pirates, cloud computing providers, amateur hackers . . . Historically, little has been known about their role in the process.

This book investigates the politics, personalities, and power that govern video format standards. The research is grounded in our joint expertise as scholars of information technology and media history, but also as practitioners and members of the media and archiving communities: Jimi has over two decades of experience in audiovisual production and preservation and spent nearly three years working for the US Library of Congress on early drafts of its MXF application specification, and Marek was a film producer before joining academia, working closely with distributors, post-production houses and archives, often both frustrated and fascinated by the idiosyncrasies of digital video formats. Our book documents how archivists negotiate the promises of digital technology and balance preservation ideals with the financial and material realities at cash-strapped institutions. It shows the audiovisual archive as a dynamic, messy space where people frequently disagree with each other, where machines often break down, but where different communities also collaborate to develop resourceful solutions to unexpected problems.

Why Archives Need Standard Formats

If the stakes for standards are stressfully high in the media production industry, they are even more unnerving in media preservation. Archivists work on much longer timelines and much tighter budgets than most media creators. In principle, an archive, museum, or library carries a duty to preserve objects indefinitely. With rapidly changing technology and a bewildering slew of media formats circulating around the world, this is an arduous task. As older digital and analog media carriers become obsolete and deteriorate, it is hard enough to ensure that the content of familiar video formats like VHS tapes and DVDs remains accessible and viewable. But imagine doing the same for obscure formats most people have never heard of, like the Capacitance Electronic Disc.[4] Obscure formats are difficult to preserve for several reasons. If a playback machine breaks, it can be impossible to repair because it has been out of production for decades, or because spare parts are not available, or because the people who knew how to operate it retired years ago. In situations like these, uniform format standards can be an immense boon to the preservation of audiovisual heritage. Preservation experts know this, but it is only recently that archives have begun developing standards of their own rather than passively accepting materials they inherit from the film and broadcast industries.

Standards are a way to ensure the interchange of audiovisual information between people, technological systems, and institutions. The chief complaint Hermann Lewetz received about his presentation in 2010 was that the FFV1 encoding had not been standardized by professional bodies like ISO, SMPTE (the Society of Motion Picture and Television Engineers) or MPEG (the Moving Picture Experts Group, a working group of the ISO). These are the traditional gatekeepers of technological standards and format specifications for the media industry, and the documents they produce tend to be heavily skewed toward the media futures anticipated by North American, European, and East Asian industry giants: media conglomerates, film studios, large broadcasting corporations, electronics and hardware manufacturers. Audiovisual archives do not traditionally have a seat at this table since they, unlike media producers, are not seen as the intended and primary users of such technologies, and media's *pasts* are often an afterthought in standards development—sometimes with dire consequences, as we will see in later chapters.

Industry-influenced standards have always been a critical part of audiovisual preservation. The perception of stability and approval from Hollywood and from the broadcast industry can make a video standard seem much more appealing to a memory institution that does not employ specialists who have a deep understanding of video engineering. Many of the interviewees in this book cited the lack of standardization by professional bodies of open-source technologies like FFV1 and Matroska as cause for concern about adopting them, especially with their use in preservation being so sparse around 2010. More than a decade later, the situation has changed significantly. It became clear to some preservationists that such open-source technologies sometimes work *better* than solutions from the media industry. But for FFV1 and Matroska to find purchase in libraries and archives, they would need to be standardized by a body that both the archivists as well as the broadcast and Hollywood industries would recognize as an authority in digital video technology.

It is this impulse that has propelled many cultural heritage institutions worldwide to join international format standardization initiatives. Among them are influential European and North American film archives and libraries whose preservation policies will shape the future of audiovisual heritage for generations to come. Individual archives have many motivations for selecting a specific format—from financial pragmatism, infrastructural or staffing necessity, ease and transparency of implementation to moral and political principles. But standards are contagious. The cultural capital that large archives and libraries wield in shaping standards creates top-down effects, because smaller archives are more likely to adopt the same practices and file formats rather than develop their own. This social character of standards can sometimes not only slow down the adoption of new tools and technology but also impede preservation endeavors at many institutions that are either unable to perform high-quality digitization in-house according to the latest standards or are taking a "wait and see" attitude until a well-adopted front-runner emerges out of the many competing technologies. But before we discuss the broader practical and theoretical ramifications of these tensions, it is worth pausing to reflect on what a format standard *is* and *does*.

What Is a Format Standard?

The term "standard," as the computer scientist and standardization expert Kai Jakobs observes, can be devilishly hard to define. Jakobs notes that while there is no universally accepted definition of what a standard *is*, there are two important elements that standards tend to *share*: They are designed by some kind of recognized and legitimated authority and they are usually established by consensus.[5] Standardization, as the media scholar Matthew Fuller specifies, implies "the necessity of relations" among institutions, technical protocols, practices, and people.[6] By ensuring interoperability, standards work as a mode of governance within media systems, encoding industrial and social relations into technical protocols.[7] From a business and economics perspective, one of the key goals of standards development is to help create a market wherein a standard solution is preferable to nonstandard solutions.[8] In other words, standards tend to be developed with the goal of insinuating them into existing markets or completely changing the markets in favor of those standards.

STANDARD POLITICS

In December 2018, the Apple company announced that it would no longer support JPEG 2000 in future releases of the editing software Final Cut Pro, one of the workhorses of the film industry upon which many filmmakers rely for their work. Nominally, the company cited its transition to strictly 64-bit computer architectures as the reason for phasing out JPEG 2000 alongside a handful of other common and popular compression codecs and formats.[9] One of the factual consequences of this strategic decision, however, is that it also imposes an incompatibility with old software. Post-production studios do have other tools to export JPEG 2000 files, which are the basis of digital cinema and an important production and storage format in the broadcast industry. But there is an argument to be made that this move by a single large actor in the hardware and software market could significantly increase the difficulty of creating and maintaining JPEG 2000 files for some people and businesses in the film and broadcast industry. It therefore also may have purchase on broadcasters' inclination to adhere to JPEG 2000 as a standard in the value chain of, for example, a TV show.

Contrarily, neither the FFV1 nor the Matroska format are supported natively in major video editing and encoding tools. Online discussion forums document that archival professionals have been demanding support for these formats for at least a decade now, and they often resort to various informal tools, plugins, or custom-made code with precarious, inconsistent, or erratic results. But in contrast to the past, archives in the twenty-first century not only passively react to the ebb and flow of formats in the media industry but have begun influencing it. The increasing adoption of FFV1 and MKV by archives could incentivize software developers to begin supporting them more widely.[10] Support in common editing tools would then make it easier for filmmakers, television studios, and post-production companies to work with these formats, too. And indeed, we are beginning to see some initial signs of this: In May 2024, the popular open-source video transcoding software Handbrake added support for the FFV1 encoder, including the new preset "Preservation FFV1."

These examples show how compression codecs and container formats can be followed like a trail of ripple effects that propagate across the media production, post-production, exhibition, and preservation life cycle. Pursuing the processes by which formats are developed, standardized, used in practice, and abandoned can allow us to address industries that are rarely considered in concert because their interrelations appear remote and indirect, but which nonetheless can have powerful effects on each other and on the culture and heritage of the moving image.

One facet that makes standards so powerful is their ability to propagate through time and space. This is, in fact, one of the goals of digital file standards—to allow users to create files in a way that ensures that playback hardware and software will be able to decode them anywhere in the world, in uniform ways and, ideally, forever. Sociologists of science frequently emphasize this "process of constructing uniformities across time and space, through the generation of agreed-upon rules."[11] Standards are, of course, a kind of encoded power in themselves.[12] A standard says, essentially, that *this thing* can only be considered *this kind of thing* if it adheres to *this* specification. Although technological standards tend to fade out of sight in our day-to-day lives,[13] it is herein that the understated weight of standards lies. Ashley Blewer, an archivist involved in standardizing one of the formats at the center of this book, confirms what scholars of technology have long been theorizing: "There's a lot of soft power in these things, and that often goes unstated. [. . .] It's politics, really. It's all political work, even though it seems like technical work."[14]

Blewer's summary concisely confirms the "intimate connection between standards and power."[15] This is one of the key theoretical tenets of science and technology studies (STS), the scientific discipline for which processes of technological and epistemic standardization are a central concern. Standards development does not usually transmit the more overt kind of political power that involves armies or economic embargoes. Standards can be dull, fussy, and hard-to-read documents, made by committees, working groups, and task forces that are rarely seen as very glamorous. But those documents create areas of influence and lines of political, cultural, or economic division. They can give certain companies or countries technological advantages in some regions of the world or regulate which businesses can access a sector of industry.[16] They can make content in a particular language or made by a particular social group unavailable or difficult to view in a certain area. They lock users into certain technological paths, making it difficult to purchase machines from a different vendor down the road.

History teems with examples of technological standards and media formats being deliberately instrumentalized in battles for market supremacy or cultural dominance, and to draw geopolitical and ideological boundaries: think of the development of the mutually incompatible NTSC, SECAM, and PAL analog television standards, the competition between the X.25 and TCP/IP Internet protocols on either front of the Atlantic, or the "format wars" between VHS and Betacam and later Blu-ray and HD DVD.[17] In many countries, the long-lasting legacies of divergent analog standards of the past still persist as a shadow of colonial influence, manifesting as visual glitches in video signals and technological incompatibilities in hardware.[18] The various subsequent digital broadcasting standards or "national" compression technologies, such as China's AVS video format,[19] echo similar technopolitical concerns about media-technological autonomy in the present.

Archival standards have seen particularly little interest from scholars of technology. At least two reasons for this seem plausible to us. Firstly, the implementation of standards in archives is downright niche in comparison to the seemingly more momentous standardization of large and ubiquitous infrastructures like the telegraph and railways and electrical and power systems,[20] network and cellular technology and the Internet,[21] education,[22] medical tools and protocols,[23] or various procedures and measurements in science and industry and government administration[24]—some of the favored subjects in academic debates on the history of technology and standardization.

Secondly, archival video, especially its newer digital forms discussed in this book, is lodged between the rather distinct worlds of media format standards on the one hand, and of information and communication technology standards on the other. Paradoxically, digital video format standards exhibit both the rapid development and inconsistent implementation that some scholars typically associate with information technology,[25] but are simultaneously bound to the longevity and consistency required in an archive and more commonly seen in traditional media standards. This awkward abeyance between two standardization cultures and several industries—film, broadcast, software, and, most appositely, preservation—means that the jurisdiction for developing standards can be uncertain. Chapter 3 will spell out how such ambiguities steep the technical work of standardization in highly politicized discussions of access and ownership. The FFV1/Matroska and JPEG 2000/MXF communities that we discuss here are similarly divided in their alignment with either an open-source ethos or a confidence in strong, influential, and respected institutions. Ultimately, these media formats and the standards that describe them are about power. They are about the freedom to choose a way of doing things or about the pressure to model one's mode of operation on that of another. Archival standards are about the negotiations of power between broadcasters and broadcast archives, between the film industry and film archives, between computer manufacturers and creative workers, and between standards bodies and archival institutions.

Scholars observe that standards are usually compromises made by a community to serve various needs, biases, and power relationships.[26] Making these relationships visible is critical to understanding the adoption and success or failure of a standard. Communication technologies often work best when they are standardized because this helps to ensure their interoperability.[27] The uniformity and predictable operation that a standard enables in the ideal case is critical for the smooth, error-free exchange of data between users and systems. Standards are usually designed to allow for interchange between other technologies. These standards are, therefore, considered *coordinative* because they "frequently aim at interoperability and compatibility of technology with the effect that transaction costs are reduced and positive externalities are generated."[28]

The social character of standards also has its downsides. As standards are used by more institutions of high prominence, other libraries and archives may adopt them either out of necessity or out of a desire to emulate other users instead of any intrinsic benefits, thinking along the lines of: "These are the standards everyone is using." This can lead to what economists and science and technology scholars call "lock-in" or "path dependence,"[29] "trajectory,"[30] or even "cumulative mess trajectory."[31] In other words, this is a situation wherein a system develops a momentum that maintains and sustains it.[32] Once the path or trajectory has been set—for example, when an archive has chosen to purchase digitization equipment that only produces files in a particular format—it can be extremely costly or impractical to change to another standard even if the "path" leads to results that are suboptimal or ineffective.

Deference to existing standards can be further problematic when communities rigidly adhere to them without question. Standards can come to "function as an alternative to expert authority—a way of embedding authority in rules and systems rather than in credentialed professionals."[33] As our history of the development of Matroska and FFV1 will show, there can be considerable resistance in a community to new standards that diverge from accepted practice. As Erwin Verbruggen from the Netherlands Institute for Sound and Vision explained to us,

> The opinions are harder to overcome than the technical aspects. Because making decisions on formats, it's ultimately a question of: what are you investing in the future? [. . .] It's a hard conversation because emotions come into play. [. . .] 'Tribalism' is the wrong word to use, but it's also a bit of: What adherents, what kind of group do you want to be a part of?[34]

At the same time, some users may adopt new technologies at least partially *because* they are new. This bifurcation is particularly noteworthy because scholars have observed it happening in completely unrelated communities, too. For instance, some developers of encrypted messaging apps also no longer trust nor comply with the output of formal standardization bodies, and "standards" instead emerge accidentally or by convention.[35] Could this be an indication that we are witnessing the slow decline of the global apparatus of formal standardization bodies, at least with respect to digital technology?

That this may be happening has been suggested already by the mid-2000s, but there is an ongoing debate among scholars about whether to understand the relations between formal, international standard-setting bodies and less formal consortia and forums (in addition to, *inter alia*, national governments) as complementary, competitive, or substitutive.[36]

Among archivists, we have found that emerging video formats are frequently framed in antagonistic terms as technologies that rival each other on an exclusionary basis, although every archive with a substantial collection is likely faced with many different video format standards and, as later chapters will show, simple dualisms like formal/informal may not always be conducive to understanding the reality of standards development nor of archival practice. Moreover, the history of standards that we tell in this book is exceptional in that it involves an unusual and sudden passage of technologies from one extreme end of the informality spectrum (political critique and piracy) into formal standardization.

There is a "sweet spot" in standards development wherein enough actors are enrolled in the process to develop a highly interoperable standard but not so many are involved that the standard fails to achieve closure and stability.[37] Jonathan Sterne's study of the MPEG, the standards-making body that formalized the MP3 as a digital audio compression format, is illustrative here. Sterne, a media and communications scholar, describes the culture of MPEG as one of mostly for-profit organizations working together (and, at times, at cross purposes) to anticipate the transition to digital moving image distribution.[38] The MP3 format standard is used in broadcast, home consumption, and information technology. Therefore, unsurprisingly, broadcast, IT, and electronics companies were all involved in its development, and they all inscribed their values into it by making it a highly efficient computer format specification that can compress and encode audio at moderate quality and low file size.

In the words of Dylan Mulvin and Sterne, working backward from a published standard by interrogating its limits "can uncover the epistemic and aesthetic politics of the engineering imaginary."[39] Much like audio, the standards for video compression and video file formats are similarly tinged with the values of the many communities that participate in their development. This includes both corporate and nonprofit actors (although that distinction is more complicated than it may seem) but it also involves unexpected new resonances between groups like archivists and film pirates. However, looking at the published standards documents themselves can only tell us so much about their politics. This is where the notion of "format" is useful to analytically complement the study of standards.

BETWEEN STANDARDIZATION AND FORMATTING

While standards have been a mainstay of research within STS and sociology, a neighboring area of inquiry is also materializing around formats. "Format theory"[40] or "format studies"[41] cuts across university departments, linking fields as diverse as media and communication studies, sound studies, film studies, book studies and literary scholarship, library and information science, and art history. More or less simultaneously, many scholars in these fields seem to be turning to the term *format* to understand, explain, and theorize contemporary and historical culture. The utility of *formats* lies in how they finetune the scale of analysis. Formats are central to understanding the experience of culture and media but also their circulation.[42] A format standard—like, for instance, the once ubiquitous 35mm motion picture film—can act as the synecdoche for an entire medium: For most of its analog history, 35mm film *was* cinema. All other formats were either "non-standard" or had to be marked by some additional qualifier. But because of this conflation, scholars have argued that an expansive term like "medium" or "cinema" "does not constitute a coherent whole [. . .], it does not represent a relevant level for analysis."[43]

Formats, on the other hand, are a smaller category of thought and experience than medium. By narrowing the focus down to this level, we can begin to address the nuances, all the other forms of the moving image that are nonstandard—the "non-cinemas" as Haidee Wasson calls them with intentional irony, given that films in substandard formats vastly outnumber those made to be shown in cinemas.[44]

Most pertinently for our purposes, formats incessantly point to the materiality of media objects.[45] Discussions of media often conceive of them as cultural, economic, or social formations in which technology serves to *mediate* interactions between social groups but does not *intervene* in them. As chapter 1 makes clear, however, standardized technology often does not behave in the way envisioned by the standard. There are important discrepancies between the standard specification as an aspirational ideal, and format as actualized, implemented form. The ubiquitous JPEG image file format illustrates this excellently. Thomas G. Lane, organizer of the informal Independent JPEG Group, once wrote in 1991: "The draft JPEG standard is an extremely loose document: many decisions are left up to the individual implementor. People working in isolation are likely to produce implementations that can't exchange JPEG-format files."[46] It can be argued that the subsequent overwhelming global success of the JPEG image format is attributable less to the formal standard specification document than to the fact that there was one free, accessible, reliable implementation: JPEGlib. It is this interpretation of the JPEG standard that was incorporated into early web browsers, and, remarkably, it had been developed by the informal Independent JPEG Group in complete autonomy from JPEG standardization bodies.

Much of this book revolves around this divide between ideal standard and actual format. Its consequences can be devastating in an audiovisual archive. To give another example, besides the public document that describes it, the Tagged Image File Format (TIFF), common in image and digital film preservation, has many private, nonstandard, undisclosed, or unreliably circulated implementations.[47] This means that two TIFF images can be structured in vastly different ways and contain several types of information. The video container formats MXF and MKV, which play two of the leading roles in this book, were plagued by many of the same problems. Such variations within the same format pose major problems in preservation because even if an archive receives files in an ostensibly standard format, the standard could have been implemented incorrectly, which can cause errors and failures—unreadable files, corrupt data, and irreversible loss of records.

In an ideal situation, therefore, "standards assure that a format that operates on one system will operate on another," as Jonathan Sterne has observed.[48] Some academic literature emphasizes how formats serve as "boundary objects," a term we will explore more fully later, and enable collaboration between different communities of knowledge.[49] But as our book documents with many examples and as some earlier literature has recognized,[50] in practice, this is not always the case. Standards are often implemented inconsistently, machines do not produce the standard output we expect them to, and network protocols glitch. This is why newer format standardization initiatives like PREFORMA, a large-scale European project involving numerous cultural heritage institutions, have begun standardizing not only the *specifications* for formats but also providing their *implementations* for reference.

Formats thus seem to vacillate between the rigidity of industrial standardization and the messiness of media practice. A media format often *is* a type of standard—"codified in some way . . ." to have a "contractual and conventional nature."[51] But not all formats are standardized. Even if they are, the standard does not always fully coincide with nor contain the multitude of formatted media objects "on the ground" in the archive or in the computer. Whereas Sterne posits that the "moment of birth for a format would have to be the moment it becomes a standard,"[52] the history of cinema (and its preservation) offers numerous examples to the contrary. About a century before JPEG, during the early years of moving picture exhibition, many inventors were competing for market dominance with various film projection formats without any formal standards to rule them. The Kinetoscopes, Bioscopes, Biographs, and other formats were all viable film projection technologies that differed in construction, film gauge, aspect ratio, and so on—they were, in other words, distinct *formats*. Yet they not only preceded the emergence of formal standards but even the idea of cinema itself.[53] FFV1 and Matroska have also existed as video formats for two decades now but are only about to become standards in the near future. In fact, when Ashley Blewer, one of the preservation experts involved in standardizing FFV1, was telling us about the compromises they have had to make, she observed:

> Something that comes to mind is having to stop and select a version of FFV1 for the standard, even though the development of FFV1 continues as part of FFmpeg development. So, we had to stick to version 3 as the official recommendation, even though there's some improved features in versions 4 and up.[54]

To scholars and practitioners interested in formats, this is fascinating because it shows how the standard and the format follow different temporalities. Intuitively, it may seem that a format developer and a standards developer do largely the same thing, when in fact their work diverges in its relationship to time and to the permanence of their product.

Based on our conversations with archivists, we, therefore, argue that a fine but meaningful distinction exists between standards and formats and that teasing out the incompatibilities between promise and practice can help us better delineate the "epistemic grounds upon which a technology may be understood."[55] In recognition of the ambiguity, we use these terms in the compound form "format standard" to point to the fact that the vision of a standardization committee never seamlessly aligns with the material reality of a formatted media object in use. Our book uses this semantic gap, which also points precisely toward the disciplinary interstice between STS and format studies, as a point of departure, drawing on both fields to understand how format standards are made, as well as how they work (or not?) in practice.

UNRULY FORMATS

Unruly formats have been troubling heritage institutions for decades. Film and video archives have been particularly vulnerable to format issues due to the complex and international nature of the moving image itself: its tendency to circulate many different "originals" that are reformatted haphazardly,[56] and the accompanying conceptual and cataloging difficulties involved in making sense of the copious possible versions, editions, and formats of a single film. Films get censored and reedited, dubbed, and subtitled, reformatted to smaller film gauges or compressed and distributed on video.[57]

As we will discuss in more detail in later chapters, media-historical developments are often followed by a cycle of proliferation and then standardization of storage formats. This holds true across media as diverse as movable print, recorded sound, photochemical film, or electronic video. Technological inventions in these fields often result in the development of a range of competing technologies which, however, are then often seen as a hindrance to the effective functioning of industries, governments, and their institutions. Standardization efforts then tend to kick in, often borne out of an administrative or economical need to rein in media format anarchy.

With increased computer usage among film archivists, format standardization became one of the urgent goals of the community already in the late 1970s and early 1980s. During this time, it was still too soon to try to standardize the congeries of film gauges, magnetic tapes, and incompatible video files that cause preservation headaches today (Fig. 0.1). Instead, archivists started by attempting to standardize the format of catalog data. The development of a common computerized cataloging format would—so it was hoped—facilitate international exchange, consistency, and discoverability, and was among the International Federation of Film Archives' (FIAF) major priorities. As the film and video archivist Roger Smither has documented, despite widespread support throughout the 1980s, such standardization initiatives often encountered difficulties not only at the level of technology but also administration, logistics, and politics. Quick progress at individual institutions can paradoxically impede international acceptance, adoption, and interoperability.[58]

Archives—audiovisual archives, but also general archives and museums—commonly encounter media objects in their custody whose format is difficult or impossible to establish. This could be, for example, an electronic file whose format is unidentifiable and whose contents are therefore illegible even though they can be read. Problems with mysterious files have led to the development of format databases that support archives in the identification of electronic objects, such as the now-discontinued *Unified Digital Format Registry*.

Today's format incompatibilities and the associated questions of standardization and reformatting are therefore not a new set of problems. The ongoing process of standardizing the "rogue" formats FFV1 and Matroska takes up a number of these decades-long concerns, but it reappears in a different industrial, institutional, cultural, and technological climate: that of digital media.

FIG. 0.1. Creating order out of chaos: A collection of different tape formats at Illinois Public Media (WILL), Urbana, Illinois, USA. (Photograph by Jimi Jones.)

The Great Upheaval: From Photochemical Film and Analog Video to Digital

For more than a century, photochemical film production, exhibition, and preservation were the essence of moving image heritage. Traditional motion picture preservation workflows and best practices have been focused on the physical object—film or videotape—stored in stable climates and containers on shelves. But in the past decade and a half, film production has almost fully shifted from photochemical to digital. The change from chemical to digital image capture has brought with it a deluge of digital moving image recordings, afforded by the cheaper costs of video production.[59] The previous approach, where the physical medium can be passively preserved through a hands-off "benign neglect" is insufficient for the needs of the file-based, digital environment. As we will see in later chapters, digital preservation must be proactive because files can be very easily lost due to technological failures like hard drive crashes or user errors like accidental erasure of data. The dizzying pace of technological developments—from rapidly increasing processing speeds and network bandwidths to growing screen sizes and color reproduction technologies—also means that moving image formats have much shorter shelf lives. A roll of 35mm film made according to the specifications standardized in the 1930s is still playable on a motion picture projector today, while hundreds of video compression codecs and formats have come and gone in just the past three decades. Such format troubles had already emerged with the introduction of analog video technology in the 1970s and 1980s, and digital video has aggravated them.

Today's motion picture film workflows look radically different from the earlier, all-photochemical and analog video world. Production companies and archivists have had to develop new skills and practices: file management, metadata harvest and generation, fixity checking, regularly migrating formats and encodings—these are the trappings of the digital IT infrastructures that dominate moving image production and preservation in the twenty-first century. In order to preserve digital material, modern archives have become very high-tech environments.[60] Moving image archivists had to transition from being stewards of physical objects to *also* maintaining *digitized* objects, as well as a ballooning volume of *born-digital* objects, and the disparate infrastructures and knowledge systems which all of them necessitate.[61] This technological change has radically reformulated what preservation means. And with it, the profession has had to redefine the work and identity of the moving image archivist, too. Reading it alongside changing format standards, this book makes visible the changing character of present-day archival work and the larger struggles for its social legitimacy: Who can be considered a professional archivist? What does digital audiovisual archiving even *mean*?

One of the earliest extended discussions of moving image preservation in the United States focused on the sociotechnical milieu of photochemical film, Anthony Slide's *Nitrate Won't Wait,* was a landmark book about the dangers inherent in the benign neglect of motion picture film, particularly the earliest cellulose nitrate films.[62] Slide's work was an important step toward opening the world of film preservation to a wider audience. While photochemical film preservation is still very important, the needs of analog and digital video must also be addressed because of its fragility and the growing obsolescence of video formats and playback equipment. Digitization has become the accepted long-term preservation strategy for this kind of material, which has further propelled the shift from the analog to the digital in moving image preservation.[63]

The idea that one could preserve audiovisual objects by separating the contents from their material carrier and turning the former into digital code seems commonplace and uncontroversial today, but this was not always the case. It required a complete shift of archival paradigms that occurred gradually though not simultaneously across different types of archival institutions. According to some accounts, sound and audiovisual archives were initially more amenable to this idea than traditional textual archives.[64] In the late 1980s and early 1990s, the notion emerged that for electronic records, "preservation means copying, not physical preservation."[65] During the 1990s, libraries and photographic archives were the first to start experimenting with the use of digitization as opposed to microfilming for preservation reformatting. The issues then were nearly identical to those faced by film archives today: obsolescence of hardware and software, proprietary and therefore opaque technology, incompatibilities between vendors, and lack of comprehensive and inter-institutional integrity verification methods.[66] Here, too, format standardization was one of the big ambitions that led to the formation of the Digital Preservation Consortium in 1994, today the Digital Library Federation.[67] In the domain of *audiovisual* preservation, efforts like these are only just starting to come to fruition with the broad adoption of international preservation assessment frameworks and ongoing large collaborative format standardization projects.

THE CONCEPTUAL FOUNDATIONS OF THE ARCHIVE ARE STIRRING

Most of the professional and academic discussion of moving image preservation focuses on the *technologies* involved in the transition from the analog to the digital. For example, in the mid-2000s, the major Hollywood film production studios invested considerable resources into studying the effects of the transition from photochemical to digital on their industry.[68] One of the key findings of this report was that as costs of digital production and post-production continue to drop, the need for digital file management becomes ever more critical. This finding was echoed by a report funded by the U.S. Library of Congress, which emphasized the need for standardized digital video wrappers (like MXF) to encapsulate metadata along with video and audio information and to encourage ease of exchange and preservation of video files.[69] This report was influenced considerably by the departments in the Library of Congress that had been working on the so-called AS-07, a document that specifies the library's implementation of MXF as a preservation format which we discuss in chapter 2.

There are very few scholarly works about the effects of the digital transition on the *meanings* of preservation, quality, and sustainability in moving image archives and libraries. In the era when "film preservation" referred to photochemical film, the concept of sustainability hinged on the conservation of the physical object—the roll of plastic film coated with emulsion in a can on a shelf. Today, concepts of fidelity to the analog originals and of what constitutes "sustainable preservation" or "high quality" with respect to the digital surrogates have become much more fluid, albeit very central to the contemporary world of file-based digital moving image preservation. Much of the work that moving image preservation professionals and developers of related standards do depends upon an ongoing negotiation of just what these terms mean in the file-based realm. The human geographer Graham Pickren, following Stefano Ponte and Peter Gibbon, argues in a different context that the capacity to institute standards is not simply a function of class power, but also reflects the power to establish a dominant narrative about what constitutes "quality."[70] Transplanting analog preservation concepts in a way that they could continue making sense in an entirely novel, digital context was a formidable task that required the redefinition of such foundational terms as "loss" or "quality." New ideas and new language have entered the archival vocabulary. Much like earlier debates surrounding the changing understanding of "permanence" or the relative benefits of emulation versus migration,[71] audiovisual archivists now routinely speak of lossless files and fixity checks—notions that would have made little sense in a "traditional" audiovisual archive as it once was. The following chapters document how these new terms and skills, practices, and workflows found their way into the archive.

One important study of the transition from celluloid to digital is Giovanna Fossati's work on the move from traditional film preservation workflows to the file-based, digital environment in the early twentieth century. Fossati discusses how this still ongoing technological change affected how film archivists see themselves, their work, and the meaning of the moving image broadly.[72] Our book looks at the critical developments that have taken place since: the increased digital storage volumes, the availability of networked and online preservation systems, and, perhaps most saliently, the development of new format standards, which create fascinating new areas of contact and conflict between different types of archives.

On the one hand, the newly developed formats compel archival institutions to take a stance on whether to adopt closed or open-source technology. This choice has profoundly reshuffled some of the traditional alliances and reached beyond *film* preservation into the much wider domain of *video* preservation. It thus invites into the conversation many other institutions besides the traditional film archive, such as libraries, university research collections, or media museums. Although some scholars object to the use of the collective term "memory institutions" to refer to organizations with vastly differing missions and practices,[73] the colossal growth of accessible media equipment has meant that even establishments that did not traditionally collect audiovisual records now have to figure out what to do with digital video. On the other hand, the design of these digital formats curiously perpetuates the logic of some of the analog media they have supplanted, in fact complicating the seemingly self-evident distinction between the analog and the digital.

HOLLYWOOD NO LONGER REIGNS SUPREME

Another crucial development that followed the transition to digital media relates to the wider relationship between media production and preservation. In the early twentieth century, motion picture film preservation in the United States was mostly a state-funded endeavor that was later tied to copyright submission. In her book on the formation of film preservation institutions, Caroline Frick argues that because Hollywood productions featured prominently in this early film preservation milieu, the American film preservation endeavor has, over the course of the twentieth century, established the narrative that these privately produced motion pictures are synonymous with the American film heritage.[74] This narrative continues to be promulgated by the Library of Congress, which focuses much of its resources on the safeguarding of Hollywood and American broadcast television productions. In Europe, too, the film industry's reliance on state funding, national broadcasters, and their commissioning editors has resulted in a deposit pipeline wherein state-funded productions are typically preserved in state film archives. The preservation of video heritage is much patchier internationally, with the Council of Europe's Protocol on the Protection of Television Productions—the first international treaty on mandatory deposits—only entering into force in 2014. Media that do not fit these criteria—like small amateur collections, research collections, private or independent channels—float somewhere in-between.

The passage from analog to digital image capture has similarly been propelled by the major players in the cinema, broadcast, and electronics industries and their devices for the creation and playback of moving image materials. In the past, libraries and archives were obliged to use machines built and marketed by electronics companies like Sony, Panasonic, and JVC to preserve photochemical film and analog video. Now, in the era of digital media production, software developers like Apple and Adobe also participate in the market in playback and transcoding instruments.

One considerable difference since the publication of studies like Frick's and Fossati's, though, has been the emergence of many tools created by and for the open-source and freeware communities. More institutions are developing their own digital preservation methods and extensions of existing software tools for their local practice. Media preservation tools in libraries and archives today often do not have as much of a hand-me-down quality as in years past. Some archival institutions are turning away from the traditional gatekeepers of audiovisual standards. The line between producers and consumers of media production and playback technologies has become blurred as more memory institutions adopt, adapt, and invent their own practices, specifications, and standards.

This is a fascinating process to follow because the reconfiguration of the relationship between media industries, standardization bodies, and archival institutions will have ramifications into the future: The video formats in an archive partly dictate how audiovisual heritage gets preserved, how easily it can be accessed and migrated, and how it can be reused. The unexpected new alliances and points of contact—such as between archives and film pirates—reveal a great deal about the larger functioning of contemporary media culture and the complex ethical, political, and technological catalysts that drive it.

THE CHALLENGES AND CONSTRAINTS OF DIGITAL AUDIOVISUAL PRESERVATION

Media preservationists—particularly those in the nonprofit cultural heritage sector—encounter many challenges in their work. Equipment and format obsolescence and fragility, technical knowledge bases that are shrinking over time, and the lack of universal best practices and target formats in digitization all plague the endeavor of audiovisual preservation and accessibility. As Peter Kaufman puts it, media preservation is a "defensive vocation" involving pitched battles with all these factors in what is, on a long enough timeline, inevitably a long defeat.[75]

Video preservation and digitization in the cultural memory context is still a relatively young endeavor, having only been performed in many libraries and archives since the turn of the twenty-first century. As the editors of the recent technical guidance document by the International Association of Sound and Audiovisual Archives, *IASA-TC 06: Guidelines for the Preservation of Video Recordings* note, "video preservation practices are not yet mature, format specifications are still being refined, and there is relatively little actual experience."[76] The lack of intimate familiarity with the highly technical nature of analog video engineering and digital formats in many memory institutions complicates or completely obfuscates decision-making with respect to digitization outputs. To make matters worse, the ongoing, unimpeded breakdown of media carriers means institutions may only have one chance to digitize an analog object before it irreparably breaks down—only one chance to "get it right" with respect to performing the best digital transfer possible.[77]

The cost of video digitization is another challenge. Analog video technology which is no longer produced or supported by its designers is notoriously cranky of maintenance and continually more costly to procure. Moreover, an institution almost certainly has limits with respect to which media formats it or the vendors it contracts can afford to support (Fig. 0.2).[78] For example, an archive may have all the technology required to digitize VHS tapes but no capacity to digitize U-Matic cassettes. Thus, the complexity and costs of video digitization can narrow the scope of the kinds of material and content to which a memory institution can maintain access. These interoperability problems are part and parcel of audiovisual preservation and are nothing new to the current digital/digitization era, as James Snyder of the Library of Congress made clear in our interview.

The cost of data storage has plummeted in the last decades. In consumer electronics, terabyte-sized storage media are now entirely unexceptional, so the price tag associated with keeping digital objects may not immediately appear as a major challenge. But it can still be potentially crippling to any small, modestly funded institution because high-quality video that conforms to long-term preservation guidelines simply results in enormous data volumes. For example, digitizing a single hour of an analog-sourced video object into high-definition, uncompressed digital video can result in a file over 600 gigabytes in size.[79] For this reason, one of the chief challenges archivists face is "how to maintain the essence of audiovisual content in the smallest and most manageable package possible."[80] Video compression is one of the tools in an archive's arsenal that can achieve this.

Compression, in media technology, refers to a wide, complicated array of techniques employed to decrease the signal bandwidth or file size of objects while, ideally, maintaining a high degree of visual or aural quality. Moving images are often compressed "lossily" in a way that irreversibly diminishes picture or sound quality. This is typical in quotidian media culture, such as in television broadcasting and online streaming, in the digital cinema, or in amateur recordings. Archivists, however, often hold themselves to much higher standards for quality. They tend to use caution

FIG. 0.2. In order to be human-readable in the future, all analog time-based media objects will have to be digitized, like this wax audio cylinder at Indiana University, Bloomington, Indiana, USA. (Photograph by Jimi Jones.)

when employing compression in the creation and preservation of digital objects because the irretrievable loss of information can result in digital objects that do not accurately represent all the qualities of the originals. However, uncompressed video files tend to be huge, unwieldy, and expensive to store. Recently, a middle way has emerged as audiovisual archivists have begun adopting what are called lossless compression techniques.

LOSSLESS COMPRESSION

Despite all the challenges it brings, digital technology has also given archivists unprecedented possibilities to network, organize, collaborate, and share knowledge. And thanks to it, audiovisual archives have recently also acquired a transformational ability: the capacity to process and store digital video losslessly, a characteristic shared by all of the formats discussed in this book. Lossless digital file formats can compress and store information without any alterations over time. Unlike the JPEG or HEIC image files or MP4 videos that your phone camera likely takes and that discard much of the visual information for the sake of small file sizes, lossless formats preserve all the original visual data. As we will see in chapter 1, in practice, this is never as straightforward as it sounds, but suffice it to say for now that the prospect of lossless storage makes the new compression formats JPEG 2000 and FFV1 very attractive to archival institutions, whose task, after all, it is to preserve objects with as little change as possible.

Despite the obvious benefits, chapter 1 will show that the notion of losslessness is fraught with ambiguities. This recent idea in moving image preservation means different things to different people. But why study lossless standards at all? As a recent addition to the archival vocabulary, lossless formats have caused nothing short of a revolution because they seem to be the panacea to decay, perhaps the greatest threat to all archives. Yet, as it turns out, losslessness rarely does simply what it promises on the "label." It takes a great deal of work—entire infrastructures, in fact—to keep digital video files from suffering losses and data damage. And that is why lossless format standards in the archive can be a great way to understand the work archivists do: the values that guide their decisions, the considerations they weigh when they carry out their duties, and the messy material reality they encounter in their daily practice.

The lossless standards of the future are still being forged. Whereas *lossy* media have received a tremendous amount of academic scrutiny,[81] the lossless paradigm is practically foreign to media scholars, and its recent large-scale use in digital moving image preservation is media-theoretically poorly understood. Losslessness, as an operative concept, is not only relevant to ongoing research on standards in media preservation in a narrow sense.[82] It also has bearing on film and media history, especially that period of media history which has yet to take place. With the widespread adoption of lossless format standards, new digital archival objects will become, to borrow from film scholar Vinzenz Hediger, the "new and improved originals" of the future.[83] We can expect that within just a few years, most commercially circulated releases of archival material available to film, television, and media studies will be based on either born-digital material in lossless formats, or analog material digitized into them. Especially those critically endangered records that need to be preserved most urgently—severely deteriorated photochemical film and video on highly volatile tape and disc formats—are now likely to be digitized into lossless formats, in theory interrupting their ongoing decay permanently. If this prospect might delight archivists, it has at the same time caused great unease among scholars who fear what it might do to our conceptions of history and memory if technology obstructs what many see as vital processes of forgetting and decay. The ramifications warrant a deeper conceptual and theoretical engagement.

As the media and preservation scholar Zack Lischer-Katz argues, "focusing on standards and their role in mediating infrastructures and institutions can identify points of tension between different parts of the preservation community, drawing attention to otherwise unseen power differentials in the field."[84] Lossless formats are a novel category among such standards and represent a critical node or, in other words, a "boundary object" through which various communities of practice negotiate their stakes in cultural heritage.

Methodology: How to Study Format Standardization

Studying processes of digital format standardization brings with it some methodological difficulties. Format standards operate across a wide spectrum of elements that can be difficult to overlook in their entirety. On the one pole lies the microscale layer that Johanna Drucker calls "distributed materiality"[85]—the detailed computational protocols that govern how data is moved, algorithmically processed, compressed, encoded, and stored on material carriers. This is the scale at which the tiny machinic activities of signal processing and compression take place. It is here that format standards become addressable as material when technological frictions and glitches crop up.

But interoperability and standardization also act on a larger scale, on the level of people and institutions. This is where social elements come into play: the interpersonal collaborations (like the one between Peter Bubestinger-Steindl and Hermann Lewetz) as well as conflicts and rivalries (like the adverse initial reactions to Lewetz's presentation of FFV1 at the AMIA conference in 2010); the institutional wheelwork (like the many large national and international standardization projects that unfold across libraries, archives, universities, and commercial parties) and cross-industry politicking (like a supplier's decision to support or not support a particular video format).

These social and material strata of format standardization operate concurrently, but there is no established methodology that can rigorously scale the complexities and interactions between them. In a way, this bifurcation between micro and macro levels of analysis has a precedent in debates within STS during the 1980s and 1990s, out of which "infrastructure studies" emerged as a sort of intermediate tier between macro-scale research on large technical systems and actor/artifact-centered approaches.[86]

Finding a similar methodological middle way that is, moreover, adjusted toward the epistemic peculiarities of moving image media and fields of practice like archiving is no trivial task. But it is an urgent one. Lischer-Katz points to this urgency when he asks:

> How does our understanding of media archives change when we start to look to the infrastructural systems, the granular materialities of codes and their magnetic inscriptions, and their entanglement with these global ecologies, networks of global distribution and ways of knowing the world? [. . .] Without opening the "black box" of the archive and tracing its infrastructural entanglements with complex micro and macro systems and processes, we risk losing our critical awareness of the epistemological, institutional and/or ecological factors that shape the ontology and epistemology of archival media.[87]

It is here that fusing approaches from STS with format studies becomes productive in opening some of the archive's black boxes: The former can visualize standardization as taking place in a social field in which different types of actors interact; the latter, drawing on materialist media theory and history, can bring into the fold media technology's physical and infrastructural role and articulate the discrepancies between discourses surrounding notions like losslessness, quality, or fidelity and the way those technological notions are enacted in material practice.

SOCIAL INFORMATICS

MXF, JPEG 2000, FFV1, and Matroska have all found places in libraries and archives for reasons beyond their technical affordances. They have moved from their niches into libraries and archives at least partially because of social factors like the influence of their developers and users as well as the ideological stances of their adopters. Social informatics (SI) is a discipline that analyzes such social aspects and effects of computerization.[88] It studies how computerization works on the organizational or cultural level,[89] and how communication technologies develop and change over time, how they affect the people who use them and how these technologies, in turn, are affected by those users.

A chief emphasis of SI scholarship is to challenge the notion of technological determinism and to show that it has little analytical utility in terms of understanding how technologies evolve and how societies respond to them.[90] Determinism is the view, shared by much of popular tech media like *Wired* magazine, that technologies influence societies more than societies influence them and that technologies have enough agency to change the society around them. Technologically deterministic models usually assume that users are rational actors who tend to make the most informed decisions about which technologies to adopt.[91] Deterministic arguments give very little insight into why people adopt technologies because these arguments tend to presume that technologies evolve following their own logic, irrespective of how users adapt them.[92]

Perspectives in both SI and media theory have questioned that assumption and worked to expose the cultural, ethical, political, racial, social, gendered, and economic forces that influence the adoption and development of technologies. Choices of technologies can be constrained by any of these factors, which can make the very idea of "rational actors" analytically suspect. Instead, SI views technologies as networks in themselves, with internal dynamics and dependencies on the environments around them even though they may look and act like black boxes.[93] The format standards discussed in this book have co-evolved with the cultures they influence. They are the result of both collaborative and antagonistic efforts in which developers, users, and technologies all influence each other.

Scholars have been studying the dramatic technological changes related to information communication technologies for decades. One of the prominent developments has been the rapid influx of desktop computers into office workplaces in the late twentieth century. The standard deterministic view of the time was that increased access to computerized processing power would make workers much more efficient. SI researchers, among others, found that in many cases the reverse was true—that continual software and hardware upgrades, changes, and their related learning curves increased worker stress, consumed a great deal of work time, and created the "productivity paradox" of decreased worker output as computer use increased.[94] We have seen a similar process unfold as lossless video formats get introduced into archival workflows. Instead of bringing about an effortless form of preservation without loss, "losslessness" tends to result in fresh new types of losses, confusion, and uncertainty. These frictions point to what television scholar Anne-Katrin Weber, following Graeme Gooday, refers to as "the social character of success"[95]: It is not always the "best" technologies that survive and succeed in the marketplace. In fact, there can be considerable disagreement over what a "best" technology is and who defines what is best.

THE SOCIAL CONSTRUCTION OF TECHNOLOGY

The idea that a technology succeeds or fails because of social reasons rather than some intrinsic qualities echoes a theoretical perspective known as the social construction of technology. Applying this lens to media and information technologies shows their development as an intricate result of variation and selection, not as a smooth, orderly progression.[96] This view is particularly effective at showing why some technologies fail and why others succeed beyond a simple economic ("this technology was cheaper") or technocentric ("this technology was technically superior") perspective. This approach has been applied to technologies as diverse as the bicycle and the gendered sociocultural milieu in which it was developed,[97] or the many different projection systems and exhibition formats of cinema and television.[98]

Technologies have interpretive flexibility. A media object "means" different things for different people and those differences are strongly influenced by social factors. These meanings can change as technologies move from niches into ubiquity and then obsolescence. In the words of Audley Genus and Anne-Marie Cole, "niches can act as spaces for experimentation protected from market selection pressures, or to enable social networks supporting radical innovations to be built up."[99] Jonathan Sterne draws from theories of social construction in his studies of the MP3 audio format and how its meaning has changed over time for different users. The MP3 was designed largely by corporate interests to encode audio tracks for digitized films. However, with the rise of file-sharing sites like Napster, the format moved from niche to mainstream because of piracy—a regime clearly at odds with the original intentions of the corporate realm.[100] In Sterne's argument, the lack of strong copyright protection in the MP3 specification reflected the MPEG standardization group's vision of users as passively consuming physical media and its failure to anticipate that listeners could also use the format to actively encode and share music.[101]

Technologies change the contexts in which they are embedded, but they are not infinitely malleable. Technologies enter social and political regimes loaded with their designers' assumptions about users and situations.[102] In fact, designers of technologies tend to envision a particular kind of user and encode that vision into the final product.[103] Because of these presumptions, there is often a narrow array of contexts in which technologies can be made to work. Langdon Winner has argued that there are often extremely limited political and social arrangements in which technologies can successfully function.[104] He gives as an example the atomic bomb, a technology that by its design requires certain clearly delineated hierarchies and bureaucratic systems for it to function as a weapon, deterrent, and a tool of empire. Although video format standards are of a much less explosive constitution, it is also possible to perform a kind of deconstructive reverse engineering on them to peer into the cultures and politics of the people they are developed and used by.

We draw heavily on the vocabulary of theories of the social construction of technology—as readers will readily recognize by our use of terms like "closure" and "interpretive flexibility"—but we recognize that this analytical framework has also faced criticisms over the decades.[105] Chiefly, for prioritizing the agency and meaning-making done by people at the expense of the equally important role of inanimate technologies and other non-human "actors" that shape social practices. Sociologists of science and philosophers of technology have developed several alternative toolboxes, the most prominent of which is probably actor-network theory.[106] We, however, reach elsewhere to complement the social construction of technology in our analysis. Scholars of film, television, sound, and digital media have also cultivated a rich understanding of the material agency of communication in general, as well as of moving images more narrowly. Within this tradition, there exists an understanding of standardization that is admittedly more vernacular than that of SI or science and technology scholars, but which can therefore also encompass a more expansive set of cultural and ideological mechanisms of norming, regulation, and representation. Such research extends the scope of thinking beyond science and technologies, opening it to the ways in which the infrastructures and materialities of cinema and television are implicated in producing "modern" subjects and in the standardization and reorganization of experience, perception, and affect. Examples are many and diverse: From Brian Larkin's analyses of media infrastructures and film and video practices in Nigeria, which oscillate between standardization and hybridity,[107] to discussions of the experience of wireless technology both in the nineteenth century and in the twenty-first,[108] to critiques of the transformative effects on sociality and freedom that social media platforms and network protocols can have,[109] to the racial, body, and gender politics of standardized test images in both visual media and computer science.[110]

What such heterogeneous works of media scholarship have in common is a dedication to the politics of both matter and representation. And this is precisely one of the major blind spots that literature on standardization suffers from. Let us take gender as an example: Although several histories of major standardization bodies like the ISO and International Telecommunication Union (ITU) have appeared in the last two decades,[111] the participation of women in standards-making is barely even mentioned. Academic interest in matters of gender and technology is anything but novel or minor,[112] yet gender in relation to *standardization* specifically is acutely understudied, paralleling the invisibility of gendered and racialized labor in mass digitization projects.[113] Although it is widely understood that women are severely underrepresented in standardization bodies and that this, as a result, has led to standardized technology that is dangerous and ineffective for large sections of the population,[114] concrete initiatives to rectify the situation are really of very recent date. ITU's first Women in Standardization Expert Group event was hosted in 2016 (the second in 2022) and facilitating the appointment of more women in leadership positions at the organization was a major focus of its quadrennial conference in 2024. The IEC established a task force on diversity in 2019; the Institute of Electrical and Electronics Engineers' Standards Association launched a focus area on Gender Equality and Social Inclusion in 2021. At the end of 2022, the United Nations Economic Commission for Europe published its Guidelines on Developing Gender-responsive Standards, a handbook intended for standards development organizations with practical guidance to reducing barriers to women's participation. While gender is not a primary focal point of this book, it is important to us to acknowledge these societal developments (and research lacunae) because they are directly tied to other forms of gatekeeping and related tensions surrounding access to and participation in archival technology. Standardization, as Nanna Bonde Thylstrup phrases it, is "a mode of capturing, conceptualizing, and configuring reality, rather than simply an economic instrument or lubricant."[115] Gender—along with the skewed ethnic and geographical realities of standards-making—is one concrete throughline in the fraught political practice of standardization.

To summarize, in our discussion, we lean on a diverse scholarly heritage. Principally, SI and the social construction of technology in alliance with media and format theory, and archival science. But conceptually, our work touches upon broader current debates about how to shape standardization in an equitable and just manner. Our approach thus responds to the recent appeals for interdisciplinary analyses of standard-making and also heeds the need to foreground how ideology and power dynamics drive standardization.[116] The media and technology scholars Niels ten Oever and Stefania Milan propose that "a theory of standardization and power needs to [. . .] [consider] actors that may not necessarily appear 'mainstream' or prominent if the process is observed from 'classical' perspectives on standardization."[117] With the combined strengths of different fields, this book offers precisely such a look at the power dynamics of standard-making from the margins and embedded in a broader landscape of media businesses, industries, and communities.

RESEARCH DESIGN AND INTERVIEW STRUCTURE

Interviews with people who design, use, and adapt FFV1, Matroska, JPEG 2000, and MXF form the backbone of this book, and they are critical to understanding why audiovisual preservation professionals are working with—or choosing not to work with—these format standards. Many social scientists view the interview as a collaboration in which the interviewer and the respondent are both active participants in a meaning-making situation.[118] Critical to this view is the notion that interviewees are not simply vessels for information and that the interview is not a mission to search for and mine that information; instead, interviews are shared endeavors and interviewees collaboratively co-construct knowledge along with the researchers.[119] This ethic of co-constructed meaning-making guided our interview process. Over several years, we have interviewed thirty-one preservation experts at many institutions in North America and Europe, in a range of formats, both on-site and online. The majority of these were semi-structured, individual conversations; a small number took place in a group setting or via email. The professional backgrounds of the respondents represented herein are diverse: from library technicians, video engineers, and media preservation experts, to CEOs and digitization vendors, standard developers, and heads of film conservation at film archives and libraries.

We spoke with these interview partners about the process of developing and using a format standard, their experiences in working with formats and lossless technologies in their digitization and preservation work, and the reasoning and compromises involved in designing, choosing, or migrating to one storage format over another. Grasping the intentions of users and developers has also been critical to understanding the technological affordances of the standards that allowed them to be used and adapted in different contexts. Our extensive conversations explored the social dynamics within many organizations and the larger audiovisual preservation profession. What became clear is that, while there is a shared general sense for what "quality" or "loss" mean to preservation professionals, not everyone interprets these terms in the same way. The interviews have provided much insight into the evolution of the standards from their prior niches (piracy, television broadcasting, cinema production) into libraries and archives. They let us see how standards are molded into shape to make them work in the archive, how automation lodges itself into a tense relationship with manual work, how archivists render failure and loss intelligible and develop strategies to control them, and how new technological developments are changing the skills needed to work in this field. The answers often pointed us to the power dynamics that influence an institution's choice and helped us see that in some cases the standards were acting as boundary objects.

Our interviews cannot represent the vast, heterogeneous, and messy media preservation landscape worldwide. During our research, we also spoke to archivists in Nigeria, Ghana, and Malawi and we have found that while a dearth of resources is a universally shared struggle, the preservation paradigms and the types of format standards discussed at organizations in Europe and North America are frequently unrealistic, inaccessible, or unsuitable to local contexts elsewhere in the world. But our interviews do provide illuminating insights into present-day technical, administrative, structural, and conceptual challenges at many European and North American audiovisual archives, and an interesting cross section of different institutional contexts: from some of the largest national film and audiovisual archives in the world that stand at the forefront of international digitization projects, to smaller and more nimble repositories affiliated with non-commercial television stations or universities.

Some of the interviews were accompanied by an opportunity to observe the preservation workflows and technologies in person. Such site visits were motivated by our wish to understand not only the *meanings* in which format standards become embedded but also their functioning in concrete social and material doing. Driving this approach was thus not so much an ethnographic impulse but rather a resonance with the theoretical and methodological principles of praxeology in the tradition established by sociologist Pierre Bourdieu. Bourdieu developed this method in engagement with the phenomenology of human experience and from his own understanding of the *bodily* involvement of actors in the world as the embodiment of *social* worlds.[120] Moreover, praxeology's "focus is on situated, observable and meaningful social occurrences that are not only performed linguistically but also by tacit bodily movements and by the agency of material artefacts."[121] As we will see, such often deeply sensuous encounters with physical machines profoundly inform archivists' work, even when it unfolds in a seemingly dematerialized, nonphysical environment of digital files and interfaces.

Our interviews were supplemented by a diverse range of other sources: informal conversations with other practitioners in the field and observations at conferences, as well as textual primary sources such as the standards documents themselves, a large number of related guidelines and recommendations, archivists' blogs and preservation case reports, online discussion forums and professional email list correspondence, promotional articles, conference publications, public-facing workflows, and best practices. The archiving community tends to generate a great deal of documentation about its work and these rich sources of information often divulge the motivations and reasoning regarding standards development and adoption, helping to piece together the "informal dimensions of governance."[122] Such documents have been especially useful for MXF and JPEG 2000 because we were not able to interview many of their developers. A portion of the histories of their development as well as their technical particularities have already been documented elsewhere,[123] but not told by their users. Some communities, like the CELLAR working group discussed at length in chapter 3, are committed to transparency and conduct their activities out in the open, using online message boards and email list conversations. Monitoring such resources and attending working group meetings has helped us understand these communities and how their participants operate.

Chapter Overview

Chapter 1, "'Lossless': The Materiality of Archival Video Standards," examines how lossless video standards and archival media operate on a technical and material level. Understanding the material aspects of emerging standards is important because much of their appeal to archival institutions stems from the new technological possibility they promise to deliver: lossless digital storage. We argue that losslessness is best understood not as a technical property, but rather a set of performed practices and social relations, distributed across a complex assemblage of technologies, machines, and institutions. Losslessness can be seen as a dynamic surface that reflects unstable and permanently shifting conceptualizations of loss, failure, and decay within archival practice, and makes visible the changing character of present-day archival labor.

How the potentials of digital standards are framed in archival discourses rarely fully corresponds to how format standards really function when implemented by people in their everyday practice, in encounters with machines that are often temperamental, unreliable, or prone to failure. By closely examining archivists' accounts of their daily work, this chapter shows that the introduction of lossless standards ironically reaffirms the archive as a profoundly messy, lossy place rife with errors and frictions. Archivists respond to these challenges with creative hacks and workarounds, and improvised fixes often turn into accepted solutions. New format standards thus oscillate between the promise of digital losslessness and the lossy material realities of archival practice. The introduction of lossless standards and workflows profoundly transforms the nature of film and video preservation work.

The two subsequent chapters depict the non-linear adoption and adaptation of the standards as they moved from their niches into widespread use in the preservation of our shared audiovisual heritage. Using extensive interviews with standards developers and implementers, we tell the histories of MXF, JPEG 2000, Matroska, and FFV1. But we study them from two opposing shores: First, by way of the traditional, "by-the-books" standardization exemplified by the JPEG 2000 compression codec and MXF container format. These two archival technologies are the result of the media industry's transition from analog to digital workflows in the 1990s and 2000s, and they both partake of the high standing conveyed by respected organizations like ISO and SMPTE. While the backing of these organizations means that these formats have some measure of legitimacy and reliability in media production workflows, that on its own does not mean that they are useful to archivists, too. Some of the technological features of these standards spotlight fundamental differences between the modus operandi of an industry primarily interested in fast technological innovation and the much more elongated temporal plane of an archive. After sketching the history of MXF and JPEG 2000, this chapter details the US Library of Congress's initiative to make these two incongruent logics work by devising its own standard.

Chapter 3 documents the history of the video formats FFV1 and Matroska and their gradual ascendance from vagrant origins into archives. These two formats are in many ways the antithesis of JPEG 2000 and MXF. Developed by curious individual programmers in a spirit of experimentation, they found use in piracy and political critique before carving out a small following among a motley international community of audiovisual archivists. Their use was initially highly contentious but as increasing numbers of broadcasters, universities, and archival institutions began warming up to them in recent years, they have become a viable alternative for preserving digital video. As case studies, they provide an unusual but especially vivid example of how informal, amateur, and even illicit technologies sometimes enter formal standardization processes. This chapter chronicles this gradual change of wind, engaging with the larger issue of free and open-source standards and the effort to turn FFV1 and Matroska into "proper" standardized video formats. Many different motives, from sober financial pragmatism to ethical ideals, drive archivists toward nonstandard technology, but open-source formats nevertheless face an uphill battle.

Jointly, these two chapters show how the political, economic, and social forces involved in standards-making by a range of institutions modulate format standards' ability to become boundary objects that are adopted and accepted in digital cinema, broadcasting, libraries, and archives. As our cross section of a wide range of different institutional and regional contexts details, standards do not emerge fully formed into any profession. Standards come from negotiations and compromises between actors. Making these compromises and negotiations visible can help deepen the understanding of why the standards we use function as they do and why they are (or are not) adopted in various institutions.

The fourth and concluding chapter serves as a discussion of the boundaries we placed on the work and the limitations imposed on it by external forces. We also discuss how we had to reframe the work in order to accommodate the porous boundary between nonprofit and for-profit actors in the world of audiovisual standards development and preservation. One of the key assertions of this book is that the video standards under our scrutiny are "boundary objects" that have meanings and utility for actors in disparate domains like academic and government archives, television broadcasters, film production companies, and technology companies. This chapter also explores the shifting power dynamics between for-profit actors like Netflix and Google and nonprofit actors like many of the archivists and librarians who have contributed to the development of the CELLAR standards. We examine the collaboration of Indiana University and the digitization company Memnon as an example of just such a for-profit/nonprofit partnership. One of the themes of this chapter is "praxis," and the discussion returns to and expands upon some of the audiovisual preservation challenges mentioned in this introductory chapter. The chapter concludes with a discussion of potential future directions for audiovisual standards research. Do standards enact a kind of colonizing power as they cross the globe and find themselves employed in archives and libraries in many different cultures? Does open-source standardization automatically mean a more inclusive and welcoming environment for women? We believe a discussion of the gendered geopolitics of standards can be a rich and important vein of inquiry in both media studies and information science. After reading this book, readers will understand the rarely theorized, intricate decision-making process that underlies standardized file formats.

1

"Lossless"

The Materiality of Archival Video Format Standards

When history is not struggling with loss,
it must struggle with plenitude.
—Jonathan Sterne (2009)

Lossiness is what separates digital from analog media. Analog media deteriorate with each use and in response to environmental conditions. The magnetic signal on a videocassette fades the tiniest bit each time it passes a magnetic head. The tape itself wears down and the layer of magnetized particles abrades slowly. Photochemical films get scratched, shrink, buckle, or break down over time. Contrarily, while the carriers of digital information deteriorate, too, the information itself can, in theory, be multiplied indefinitely without loss or degradation because a digital file consists of no more than a string of numbers. Those numbers can be transferred to a new storage medium when the old one starts to deteriorate. This is a simplified (and, as we will see, simplistic) view of digital media, but it is this potential of immutability that shapes many of the discourses surrounding the archival video standards JPEG 2000 and FFV1.

Losslessness is as much an imagined cultural horizon as it is a computational feature. In fact, in this chapter, we propose that it is more accurate not to think of losslessness as a technical category at all, as it is commonly understood when dealing with digital media. On the surface, losslessness appears like a simple attribute, something a format or compression algorithm "has" or can "do." We argue that losslessness is rather a set of performed practices and social relations, distributed across a complex assemblage of technologies, format standards, machines, and institutions. Reformulated in this way, losslessness can be seen as a dynamic surface that reflects unstable and permanently shifting conceptualizations of loss, failure, and decay within archival practice, and makes visible the changing character of present-day digital archival labor. But to begin with, it is worth asking where the idea of losslessness came from and how it entered our archives. Although lossless digital format standards are a decidedly recent technology, the genealogies of losslessness as a concept are surprisingly venerable.

A Very Brief Cultural History of Losslessness

The term "lossless" is used in a range of scientific and industrial fields, but patents and engineering literature reveal that it became particularly associated with electrical engineering and physics in the early twentieth century. The notion of losslessness frequently appears in the study of, for example, capacitors and electrical transmission. A "lossless line" is the engineering term for a medium without resistance and dielectric loss, for example, an electrical cable.[1] A wave traveling along a lossless line will retain its magnitude across the entire distance. Devices like that are rare in real life (superconductors, for instance) but physical transmission lines often behave close enough to this ideal theoretical construct to be approximated as such. Considering them lossless simply makes mathematical calculations easier.

It is from this association with processes of electrical transmission that the notion of losslessness entered the vocabulary of signal processing, early computation theory, and data compression. One widespread method of losslessly compressing data was developed by the electrical engineer David Huffman in 1952. Huffman coding, as it is now known, is implemented in a large number of popular image compression codecs, including FFV1. Huffman begins one of his seminal papers with this definition of losslessness:

> An information-lossless transducer is, roughly, one for which a knowledge of the output sequence of symbols is sufficient for the determination of the corresponding sequence of input symbols. Such transducers find application in the preparation of data for transmission through channels in which secrecy is important or in which the signals are subject to man-made or natural noise.[2]

Huffman was conducting his research at MIT with support from the US Army, Air Force, and Navy during a particularly tense period of the Cold War. Cryptographic applications for lossless compression were, naturally, an essential consideration in the development of his technique. To say that a process is "lossless" means that it compresses information—such as a digital video clip, for example—in such a way that it can be perfectly restored to its original state without any alterations whatsoever. One could call this form of compression reversible or invertible, and, in fact, all these terms are used in computer science.[3] This contrasts with the lossy compression utilized in most consumer media formats, such as JPEG images or MP3 audio files. Once compressed, lossy formats irrevocably discard some information that is deemed inessential, prioritizing file size savings and lower bandwidth over image or sound quality and fidelity.

The cultural and historical significance of the adjective "lossless" merits some thought. Practices of losslessness reflect a distinctly economic worldview of communication in which processes of signal transmission are by necessity transactional: they begin and end with a form of bookkeeping, a balance of losses and gains. Archives inherited this understanding of losslessness from the field of digital signal processing, but such fiscal logics have had a long, enduring presence within the history of format standards. A prominent example are the international format standardization initiatives of the early twentieth century led by Wilhelm Ostwald. Ostwald was a renowned German chemist turned philosopher whose Nobel Prize money financed his advocacy for global format standardization. At the core of these activities was the desire to standardize paper sizes. Universal paper formats, it was hoped, would stimulate further universal standards in packaging, furniture, architecture, and engineering.[4] But Ostwald's notion of a "World Format" encompassed much more than that. He proposed a standardized format for scientific writing and for the international, cooperative, and collective organization of scientific labor at large, at the service of energy efficiency.[5] Ostwald and his partners' efforts ultimately inspired the ISO 216 paper size standards in global use today, in many ways the precursors to our present-day thinking about the utility of standards.

Crucially, already in the 1920s, a vision of economic losslessness was operating at the heart of these initiatives. Ostwald advocated for the global standardization of the $1{:}\sqrt{2}$ paper format because this ratio allows a large sheet to be reformatted into smaller pieces of the same shape without waste—that is, without loss.[6] Because losslessness was conceived of as the opposite of wastefulness in the context of early twentieth-century format standardization, it could be made productive as a sensible rationale that tied in with cultural values of frugal and responsible resource management.

Historians of paper have shown that the $1{:}\sqrt{2}$ format had been called upon for the same cultural and economic reasons multiple times throughout history. It had also been mandated by law in France in 1789 because it facilitated standardized taxation brackets (corresponding to paper size) and conveniently suited the rationalist French Republican cosmology centered around metric standardization.[7] Some two centuries later, the American physicist Albert Allen Bartlett denounced as "exceedingly wasteful" the United States' refusal to adhere to international paper format standards and the resulting complications in the manufacture of administrative media like folders and fax machines.[8] When the US Committee on the Simplification of Paper Sizes was appointed to work with the Bureau of Standards in 1920, this happened under the auspices of President Herbert Hoover's program for the "Elimination of Waste in Industry."[9]

What is instructive for the study of format standardization is thus the observation that even in very distant historical contexts and in media as remote as digital video and paper, the efforts to homogenize formats are persistently accompanied not just by aesthetic or technological considerations, but also by cultural figurations of losslessness that pertain to economic governance, fiscal policy, accounting, and parsimonious resource management.

Moving image archivists, whose work is almost always constrained by financial, material, and spatial resources, tend to be highly cognizant of the wastefulness of certain preservation formats and procedures. As a case in point: A 2012 FIAF white paper begrudgingly acknowledges that it is advisable to digitize black-and-white film into 3-channel DPX files, saving the gray photochemical film values as combinations of red, green, and blue digital pixels for reasons of compatibility, even though a single-channel monochrome scan would result in much smaller file sizes and would thus be considerably less wasteful.[10]

In the field of digital video preservation, the notion of losslessness takes on a very contemporary, mathematical inflection in the sense that "losses" become numerically measurable and algorithmically verifiable. But the same old imperative of not letting anything go to waste remains in effect.

BASIC PRINCIPLES OF LOSSLESS COMPRESSION

Losses can take many forms in an archive. The archivist's traditional foes have been environmental: flood, fire, fungus, or war.[11] But losses can also take on the more prosaic form of misplacement and mislabeling. Anne Gant of the EYE Film Museum, the national film archive of the Netherlands, gives a tangible example: "One of the easiest ways to lose something is simply to not identify it or to have it misidentified. Which means it's not literally lost, but it's as good as lost. Because if you just have the wrong number or the wrong name, it's unfindable." Gant continues:

> For instance, if you say "oh, I know that this film had a DAT tape that was found." But maybe the people who were registering didn't know what

> a DAT tape looked like, so they registered it as DVD. Then you would look and look in the files and you say, "I know there's a DAT tape here, I see a DVD, I see a negative, I see a positive." And then the only way you would find that is to call back all the information, even the stuff that you think "oh, this doesn't count," and open up a box and say—"oh, inside this film can is a DAT tape. This is what I was looking for."[12]

In a film archive, preventing this type of loss—in other words, ensuring a somewhat more conventional form of "losslessness"—thus depends not only on the archivist's ability to correctly identify and label a format but also on her memory. This is, in fact, still a very common way of archiving. Many smaller archives around the world do not have catalogs for digital files because the archivist *is* the catalog. There are various reasons for this, ranging from structural to banal: for example, the software for the LTO tape robot might not work properly or the hardware supplier's customer service is poor.[13]

However, when dealing with digital files, losslessness takes on a much more specific, mathematical meaning. A digital video can be understood as a long sequence of numbers. To say that a video has been compressed losslessly means that the file has been transformed to become smaller than the original signal, but no information has been lost in the process. That is, the compressed copy should be decodable or "decompressible" into exactly the same matrix of pixels as the original sequence. This is possible thanks to algorithmic techniques that cleverly take advantage of, remove, and encode redundancies in signals.

We can contrast this procedure with the lossy forms of compression usually found in consumer and distribution media formats—in the cinema, in television broadcasting, and on streaming platforms. There, a large amount of information loss is accepted in exchange for savings in file size, bandwidth cost, and transmission speed. Because the human sensorium is not very responsive to fine changes in color and texture, especially when they are moving, our digital devices can play back videos with greatly reduced quality and leave us none the wiser. But in an archive, this is usually not acceptable.

A lossless format thus preserves a file in a form mathematically identical to its original and theoretically allows archivists to reformat and compress video without loss of information. This quality has enormous significance for cultural heritage institutions because it seems to upend one of the greatest and, until recently, inevitable burdens of audiovisual preservation: decay. For decades, it had been an accepted "occupational hazard" of preservation that when a medium needs to be reformatted, some amount of quality loss will be inevitable. For example, a photochemical film print that had to be optically printed to a new reel because it was badly damaged would necessarily have a somewhat lower resolution than its original. The loss of quality might be minuscule but not negligible. Importantly, though, this was largely true of both analog and digital media. In theory, all digital files can be copied exactly but until only a few years ago, most digital video formats that were practically usable in a preservation setting also utilized destructive forms of compression, which compound information losses on each reformatting. It is only with the recent emergence of lossless format standards that this situation began to change.

Employing lossless compression, as many of our interviewees reported, is usually a decision made for economic reasons. Compressed video requires less physical and digital storage and the savings, especially for institutions with large collections, can be considerable. While the offsetting of storage costs is the chief reason for the adoption and spread of compression technologies, it is not without risk. Even if they survive compression unchanged, digital files can suffer damage and data errors due to failures in a storage medium, stray magnetic fields, or even freak events like cosmic rays that can alter the sequence of bits. When such incidental glitches and bit flips happen in a compressed state, they can augment errors, causing significant picture or sound dropouts in playback regardless of whether the compression is lossless or lossy.[14] Furthermore, the wavelet compression employed in many lossless compression schemes does not lend itself to being edited natively in tools like Final Cut Pro or Adobe Premiere. Therefore, errors in a compressed file can render it unusable and unfixable, as Brian Wheeler from the Indiana University Libraries pointed out to us.

Error-correcting techniques are used to counter this, but a conservative approach to the care for digital archival objects mandates not using any compression at all (Table 1.1).[15] And indeed, many archives opt to conduct their work with uncompressed formats: TIFF (Tagged Image File Format) and BWF (Broadcast Wave Format) are common for still images and audio. Sound and photographic archives have been using these lossless formats for much longer than film and broadcast archives, and therefore look back on a longer tradition of conceptually defining and maintaining "losslessness" in their practice. But handling data losslessly is computationally expensive and video files are a different ball game altogether. Video consumes incommensurably more data than sound or still images: after all, a single feature-length film consists of 130,000 or so still images, more than the entire collections of many photographic archives. Because of this, the lossless digital storage of large quantities of video material has become viable only in the last few years. And it is also only in the last

Table 1.1. The benefits and disadvantages of uncompressed, lossless, and lossy media storage

Uncompressed	**Lossless Compression**	**Lossy Compression**
Preserves original image/ sound quality	Preserves original image/ sound quality	Permanent loss of some image/sound information
Most error-resistant and robust against failure	Less resistant to errors	Least resistant to errors and glitches
Considered most suitable for long-term preservation	Considered a suitable middle-ground for long-term preservation	Not considered suitable for long-term preservation
Uncommon in video preservation	Common in photographic and sound archives, less common in video preservation	Used in broadcasting, digital cinema, video streaming and all consumer video formats. Used to be the only practicable option in video preservation.
Enormous file sizes: A 90-minute film in high definition ≈ 806 GB	More manageable file sizes: 90-minute film ≈ 268 GB	Compact to minuscule file sizes: 90-minute film ≈ 48 GB in DCI standard, 3,7 GB on Netflix, 1,8 GB when compressed with HEVC
Example codecs: YUV2, V308	Example codecs: FFV1, JPEG 2000, HuffYUV, Lagarith	Example codecs: MPEG-2, H.264, HEVC, AV1, lossy JPEG 2000

decade or so that we have seen lossless compression formats emerge in archival environments.

Due to the much higher complexity, storage needs, and computational requirements of film and video, many of these operative concepts must be negotiated anew. The file sizes of *uncompressed* video files can be enormous, and their storage remains costly for any institution with substantial volumes of material, and prohibitive for many smaller archives.[16] Yet, at the same time, the processing power of common hardware has now reached a level where *losslessly* compressing and handling digital assets can significantly reduce storage and is also feasible for many institutions. Lossless compression standards therefore fill an important niche and have staked out a claim in the archival ecosystem. They strike a balance between large and unwieldy uncompressed digital objects and small ones that are easy to store but potentially lose important quality characteristics due to lossy compression. The meaning of lossless technologies like FFV1 and JPEG 2000 is, for many who work in preservation, tied to notions of savings—of digital or physical space and, by extension, of money.

Losslessness as Technological Promise and Symbolical Horizon

In archival discourse, losslessness at first appears like a straightforward and easily verifiable attribute of digital files. Again, a digital file is a sequence of numbers. If this file is reformatted in some way, the numbers that define its salient archival characteristics will have either remained the same, in which case the reformatting can be considered lossless, or they will have changed, in which case we can say that the file has sustained losses.

In our interviews, we have found that nearly all of the archival experts considered "quality" to be such a mathematically informed metric: The size of a file in bytes, the number of lines of resolution, and the kind of brightness and color information encoded in the file told them whether it could be considered "high quality." But in practice, losslessness is much more complex, emerging out of the synchronous operations of many actors, preconditioned upon numerous material contingencies, and often functioning more like an unattainable horizon that structures and continuously reshapes archival work.

The media preservation scholar Zach Lischer-Katz proposes that standards can be examined at the level of documents, technologies, and practices.[17] Our conversations with archival experts indicate that all three can at times be at odds with each other. How the potentials of losslessness are framed in archival discourses—that is, in the documents and guidelines that inform archival work—rarely fully corresponds to how lossless format standards actually function when implemented by people in their everyday practice, in encounters with machines that are often temperamental, unreliable, or prone to failure. The *idea* of losslessness suggests a perfectly frictionless transfer of information, an ideal medium like the one invented by electrical engineers to simplify calculations. But the actual presence of lossless standards ironically reaffirms the archive as a profoundly material, messy, living place rife with losses, errors, and frictions. Grasping this difference between promise and practice is an essential step in "understanding the epistemic conditions within which digitization standards develop."[18] A focus on lossless format standards reveals not only "an ethnography of everyday life" in the archive,[19] as we might say with Susan Leigh Star, but also the changing face of archival labor: the new skills that archivists are acquiring, the new challenges they are facing, and the new solutions they are finding.

The usefulness of lossless formats in the archive seems obvious. They promise the perfect reproducibility of digital objects. But, following Haidee Wasson's suggestion that format theory should "interrogat[e] what seems natural about our mediated worlds,"[20] we approach this obviousness with suspicion and ask: what does losslessness really mean to preservation practitioners? How do archivists conceptualize loss in their quotidian practice? How do lossless workflows transform archivists' work? Our claim is that the common understanding of losslessness as a numerical quality obfuscates both its social character and the material difficulties it involves in concrete preservation practice.

At a workshop on open-source preservation a few years ago, Peter Bubestinger-Steindl, an archivist at the Austrian Mediathek (the national sound and video archive) and one of the current FFV1 developers, announced that archival objects could be preserved in "virtual immortality," since with digital media "we don't have to stick to the limitations of physical objects."[21] This ambitious pronouncement sounds somewhat out of place in the usually rather dry ambiance of a symposium on preservation. Archival experts like Bubestinger-Steindl are, of course, aware that the temptation of a digital transcendence of physical boundaries only holds under some very narrow boundary conditions: if we disregard the fact that digital code is always inscribed on physical objects, or, to put it with Katherine Hayles's classic, that "for information to exist, it must always be instantiated in a medium."[22] As a matter of fact, it is this very physicality of the archive that persistently reinscribes itself into the new forms of work that lossless format standards such as MKV, JPEG2000, and FFV1 bring about.

Notions of fidelity, permanence, and immutability have surrounded many other historical storage media, such as books printed with movable type, the telegraph, or "indestructible" Phonograph cylinders.[23] We could situate the suggestion that cultural heritage could be preserved in virtual immortality alongside long-lived historical narratives of media-technological domination over time, death, and decay.[24] This idea also recalls those entrenched Manichean cosmologies promulgated through information theory, cybernetics, and popular science fiction that separate information from flesh, and that feminist thinkers like Hayles worked to dismantle decades ago. But in practice, digital media seldom live up to the promise of lossless transfer. We all know that digital devices fail and glitch. Profusely. Information often gets lost unintentionally and erased irreversibly. Loss, erasure, and forgetting are, in fact, central to the functioning of digital media culture—if all media were lossless, the overwhelming amount of data would be impossible to transmit and store. It is this contradiction that leads the media scholar Matthew Kirschenbaum to observe that "a digital environment is an abstract projection supported and sustained by its capacity to propagate the illusion (or call it a working model) of immaterial behavior: identification without ambiguity, transmission without loss, repetition without originality."[25]

New format standards oscillate between this illusion—or call it a promise, or a working model—of digital losslessness and the material realities of archival praxis. Jean-François Blanchette and others have cogently historicized the trope of immateriality in digital computing and pointed out how it depends, paradoxically, precisely on the material conditions of computation: the spatial distribution of bits in storage media, the architecture of processor hardware, signal processing techniques like error correction, manufacturing methods like cable shielding and economic constraints like the cost of digging up roads for cabling.[26] By closely examining archivists' accounts of their daily work with lossless formats, our work similarly shows that even with digital media, the limitations of physical objects do get in the way rather often. One of our more counterintuitive findings is that the introduction of lossless workflows produces many new types of loss. In doing so, it also profoundly transforms the kinds of tasks that archivists take upon themselves.

Archivists' attitudes toward digitization and digital media have changed dramatically in the last twenty years. The initial skepticism toward digital formats in the 1990s first turned into acceptance, then a fatalistic sense of inevitability. In 1999, the historian Abby Smith still warned that "because digitization is a type of reformatting, like microfilming, it is often confused with preservation microfilming [. . .]. Digital imaging is not preservation, however. Much is gained by digitizing, but permanence and authenticity, at this juncture of technological development, are not among those gains."[27] Smith's admonition stands in stark contrast to the recommendations on Handling and Storage of Audio and Video Carriers published fifteen years later, in which the International Association of Sound and Audiovisual Archives (IASA) decisively states: "Long-term preservation of audio (and implicitly also for video) can only be achieved by converting contents into files, and by maintaining these files like any other computer data." And further: "In order to avoid copying losses, copying has to be done in the digital domain. Analogue contents have, therefore, to be digitized and, together with pre-IT digital contents, converted into files."[28]

As some curators and preservationists have already speculated, digitization in itself is not a permanent preservation solution, and other modalities, including a return to analog storage, might emerge in the future.[29] Often, and as we will see in more detail in a moment, digitization itself creates new sets of problems. Anne Gant from EYE Film Museum observes that

> we've kind of created a problem for ourselves with all those digitizations, of course. Because the film in the cells will last. You know we have these

> beautiful vaults. They're now in excellent humidity, controlled, and in excellent conditions. In theory, they should last in the vault, as films, for a hundred years or more. And of course, then if we make a digital file, it's on an LTO tape and we've created a lot of work for ourselves. Cause now we have to for sure migrate that tape in five years and migrate, and migrate and migrate and never stop. So it's kind of a . . . we have to strike a balance between scanning as much as we can and not scanning too much that we've just created this enormous digital backlog.[30]

But at the moment, digitization appears like the only logical and, indeed, inevitable horizon of action for many archives, precisely because of the promise of losslessness that it offers. But the notion of "loss" that lossless format standards refer to, we argue, accounts neither for the vast typology of damage, failure, and decay that archivists encounter and recognize in their practice, nor for the pragmatic difficulty involved in actually keeping a file "lossless."

It is worth contrasting Peter Bubestinger-Steindl's hopeful view of digital immortality with a very different account from another preservation expert. When asked to reflect on the differences between analog and digital film preservation, Anne Gant provided us with a somewhat unexpected take:

> The way that you really take good care of an object is to put in the right conditions, and it's best not to touch it. The painting would like to *not* have a thousand people in front of it every day with their dirty skin cells and breathing on it. And the fabric would also not, and films would also not, actually. They like to be in the vault, in the right temperature and if you didn't touch them for a hundred years, that might be better. And that is the total opposite with digital. [. . .] I think the digital things need to be touched instantly and need to be touched really often.[31]

Gant's statement affirms that even in the absence of direct physical contact in the conventional sense, the body's sense of tactility remains an important point of reference for archival work, even in a digital context. Her remark demonstrates that many archivists remain committed to conceptualizing their practice in corporeal terms, phrasing their encounters with digital objects via their sensory habit of "touching" them.

As the film and media scholar Mary Ann Doane pointed out in 2007, the modern conception of digitality tends to repress the sense of touch, even as the term constantly hides in plain sight the digitus, the preeminent "organ of touch, of contact, of sensation, of connection with the concrete."[32] Anne Gant spontaneously reclaims this history of touching latently present in all digital media. The "touching," here, is a stand-in for the new routines of handling that digital media require: "It's so resource-intensive because the way you would really take good care of a digital object is to spin it up, if it's on a disk, often, or just call it back and bring it back through a checksum often, right?"[33]

Unlike photochemical film, the lossiness of digital storage media is best kept at bay if they are continuously "touched" algorithmically, that is, verified often and regularly, with new archival techniques such as checksumming, which are being introduced into the archive because of lossless format standards and which we will discuss shortly. The tactile poetics of digital materiality that emerged in the interview with Anne Gant is not the conventional way to speak of digital preservation, certainly not in the many archiving guidelines where the monotonous, rational voice of management and quality control dominates. But it is a very productive way to begin thinking about the materiality of standards in digital film and video preservation. It underscores the fact that digital archival objects—both digitized and born-digital material—are not exempted from considerations of physicality and corporeality. Even if they are hidden out of sight and not tangible in the same way a film roll in a can or a VHS tape is, archived video files are still located somewhere on a drive in a server farm or in a tape robot that can fail. Hard drives and LTO tapes (Fig. 1.1) have their own ways of reacting to the world and decaying which can, occasionally, also be smelled or heard. There are, for example, reports from archival practitioners that document how small humidifiers keep their LTO-3 tape recorders (another lossless storage and compression format) company in the wintertime due to the latter's sensitivity to low humidity.[34] The machines and storage media respond to their environments. On the level of practice, digital files still need to be touched, caressed, and pampered. And, as we will show, they can suffer losses even in entirely "lossless" digital workflows.

A PRAXEOLOGY OF LOSSLESSNESS

To perform losslessness correctly, archival institutions depend on a complex, dynamic, and discontinuous assemblage of many elements. These interact and affect the conditions upon which lossless files are predicated in specific situations. Here, we will contrast the discursive promise of losslessness with the material realities of archival work to establish how archivists conceptualize loss and how losslessness figures into their practice.

In moving image preservation, material failures inform and irritate a continuously evolving web of practices in which the knowledge regarding what even constitutes a loss is subject to constant uncertainty and

FIG. 1.1. An example of the seventh generation of the LTO (Linear Tape Open) format commonly used for long-term lossless storage of digital data. (Photograph by Jimi Jones.)

negotiation. Dave Rice from CUNY TV, a non-commercial educational television station affiliated with the City University of New York, provides a particularly revealing example:

> I remember when we were starting to migrate our data from LTO-5 to LTO-7, we still had our LTO-5 reader and we bought LTO-7 readers and writers. [. . .] We found that the LTO-7 drive, when it would read the data from the LTO-5 tape, would just find a bunch of checksum errors. The data would not be identical to the data we thought we had written. But then we would read the files that failed a second time and they would be okay. And we'd start to conclude that the data did not have an error as it was written but as it's being read. [. . .] We had bought four LTO-7 decks from this company and we ended up sending them all back for different types of mechanical or digital failure. When we got replacements, that would be better, but it made us really worried about how many people are buying LTO decks without really investigating the quality of their

> work. [. . .] So, when we write LTO data now, every file is put into a batch, it's checksummed in a digital forensics XML document and then moved onto the LTO tape, and then the entire contents of the LTO tape are read back and piped into just a checksum algorithm. [. . .] This adds many, many more hours on to the process.[35]

And thus, even if the technological conditions for lossless storage are in place, the materiality of capricious machines remains unrelenting. The very complexity of digital preservation workflows itself becomes the source of new errors, generating new conditions for loss and failure and, as a result, provoking further preventive initiatives, standards, and practices.

Despite the promise of perfect transmission and streamlined, standardized workflows, the work of making losslessness work is a rather messy making-do. In a concrete sense, this is the productive power of loss and failure: They instigate new techniques, new knowledge, and new collaborations. In Dave Rice's example, an instance of failure in the machine handling the LTO tape set into motion an investigation that first needed to locate the origin of the failure. Does the error happen during writing? During reading? Somewhere else? This process concluded with the institution of a complex new system of data gathering and forensic comparison consisting of several additional steps, all of which are, potentially, vulnerable to further failure. The introduction of a lossless format did not immediately result in a preservation workflow free of frictions. Rather, it reconfigured the relations between archivists and their machines, shifting their attention toward novel forms of failure, error, and loss.

Even ostensibly lossless tools can thus be unreliably lossy. As mentioned previously, lossless files are also not immune to distortions and bit errors due to noise, electromagnetic interference, and other types of corruption. Mechanical failures of storage drives are quite common; files can get damaged by network connectivity issues or when a film lab is transferring files to the archive through FTP with the wrong transfer type. Or the software used for the splitting of video files mysteriously gets tripped up by rapid brightness changes in a video scene, or is inexplicably unable to create files with more than a certain number of frames.[36] As Franz Pavuza from Vienna's Phonogrammarchiv candidly admits in a report on the migration of archival video footage, "[t]o this day we do not know the reason for this behavior."[37]

These mistakes, mix-ups, and mysteries are unique to digital and networked environments. Archivists respond to them dynamically in various ways. Sometimes, the solution is as banal as rebooting the computer. At other times, it requires specialized computational and forensic techniques like checksumming and hashing algorithms to continuously verify textual integrity and identity, which increases the complexity of the whole exercise. A crucial distinguishing feature of these new modalities of loss that accompany lossless digital standards is that they are frequently imperceptible to humans, and therefore first need to be made visible technologically and discursively before their presence or absence can be verified at all.[38] The algorithmic measurement of loss can in itself be inconclusive since different measurement tools occasionally produce contradictory results.[39] And so, although the epistemic responsibility for comparing originals with copies is increasingly delegated from human archivists to automated processes in digital preservation, uncertainties are not eliminated from the system, only displaced. Anne Gant described to us a similar sense of unease that goes hand in hand with automating digital preservation:

> [W]hen the tapes are on a shelf and [. . .] you can actually see the tape, I'm in charge of whether it's there or not. But then there's this unsettling moment when you move into a tape robot [Fig. 1.2] and everything's just being sorted kind of randomly and it's managed by a system that . . . you really hope that system is doing what it's supposed to do.[40]

In the archive, the tension between the promise of losslessness and an ever-present material risk of failure is never fully resolved. We might intuitively assume that with increasing standardization of file formats and workflows, archives are gradually moving toward increasingly automated verification. But in concrete practice, this process is far from linear and does not necessarily develop in a single direction from manual to increasingly automatic. The Netherlands Institute for Sound and Vision is a good example of the counterintuitive ways in which archival practices can change in response to institutional developments. Sound and Vision is one of the largest audiovisual archives in Europe with a dual mission as national repository of Dutch audiovisual heritage and corporate archive for Dutch public broadcasters. Given its alignment with the broadcast industry, in the past, its traditional collection policy was strictly centered around broadcast material. The choice of format standards reflects this: Most of the institute's collection is digitized, and it follows SMPTE guidelines and generally collects and normalizes files encoded with the lossy Sony XDCAM codec wrapped in MXF containers.[41] But, as Erwin Verbruggen explained to us, with its recent transformation into an institute for media history and the incorporation of a previous press museum, Sound and Vision's collection policy has expanded and diversified. The institute is now looking for ways to collect web video, interactive formats, games,

FIG. 1.2. The inside of an LTO robot at Indiana University, Bloomington, Indiana, USA. (Photograph by Jimi Jones.)

and other media. This is an archive that had already excluded manual intervention (and thus human errors) from its ingest and preservation workflows as much as possible. But as it begins to accommodate new formats, it might have to face the need to increase manual handling of the files again. As Verbruggen summarizes, "we're starting to welcome more manual errors into our midst."[42]

The institutional circumstance of Sound and Vision merging with another archive significantly affected both the format standards in its collection and its ability to control their losslessness. Losslessness seems to rest not simply in the computational characteristics defined by the format standard but must also be sought in institutional and political relationships.

THE MANY FORMS OF LOSS

Losslessness anchors debates about archival standards much in the same way that the concept of "resolution" drove debates about high-definition television. Like resolution, losslessness seems much easier to define than it is. Despite its deceptively straightforward philological sense of "textual identity," losslessness is not an absolute category but requires further qualification. Compression codecs may produce files that are "visually lossless" or "nearly mathematically lossless." Such forms of compression preserve an identical-*looking* copy of the image, but discard some redundant information, possibly deferring degradations into the future. Because of this, they are considered unsuitable for preservation purposes. Only an arithmetically identical, "mathematically lossless" file whose bitstream has a one-to-one correspondence to some original counts as appropriate for long-term storage.

James Snyder, the head of video engineering at the Library of Congress, one of the largest memory institutions to implement mathematically lossless JPEG 2000 for video preservation, explains that choice thusly:

> We were on the bleeding edge because folks like Carl [Fleischhauer] and Jim Lindner[43] and a couple of others made a brave decision to say JPEG 2000 mathematically lossless wrapper was the way they wanted to go. And Jim Linder created the team that created the first encoder. That was exceedingly brave and about as cutting edge as you can be. [. . .] He saw a problem, and he said, I am willing to take it on. [. . .] JPEG 2000 was the first compression scheme that had a mathematically lossless—which means [. . .] completely reversible—compression scheme that left no footprint.
>
> That was remarkable, there was no other compression scheme at that point that could do that. And so that was revolutionary, it was one of the revolutionary parts of this business. Well the business since then has caught up, we've learned the lessons of SAMMA and other folks have created other things like FFV1 and Matroska. Other institutions like NBC Universal and 20th Century Fox have adopted JPEG 2000 lossless because they see the value of it. And also JPEG 2000, both lossless and lossy, has gained a lot more traction in the industry. And so, there are quite a few product lines out there which may not be mathematically lossless but they're the next steps down. Since the decision was made here to do JPEG 2000 mathematically lossless, the digital cinema

> industry has settled on irreversible 9/7 which is the first lossy version of JPEG 2000, it's a hierarchy of lossiness. Mathematically lossless is up here and then there's all the lossy formats. and so all your digital cinema stuff is now all JPEG 2000 irreversible 9/7.[44]

Losslessness was therefore a major selling point for JPEG 2000 as an archival standard, but the format offers possibilities for both lossless and lossy compression. Furthermore, the losslessness of a particular encoder is contingent on a large number of characteristics that such standards provide. Dave Rice explains that

> analog video has this very particular combination of chroma subsampling, bit depth and other pixel characteristics that reduce the [digitization and lossless storage] options pretty quickly. Even though there are maybe 20 different lossless audiovisual codecs, once you consider all the characteristics of the data produced by digitizing analog video in a way where you preserve all the original characteristics, then your options reduce quite quickly to FFV1 and JPEG 2000 and maybe HuffYUV, which now can do it, as well.[45]

The notions of losslessness and copiability thus seem, on the one hand, discursively grounded in the hypothetical format-independence of digital data—recall IASA's recommendation to treat video "like any other computer data." Yet, on the other hand, in any concrete situation, formats' material constraints very quickly reinscribe themselves into the process to remind us that there are no data without a format. For example, FFV1 video files and DPX (Digital Picture Exchange format) images are both "lossless," but they are lossless in very divergent ways because their heritage traces back to entirely different habits of media practice. Film archives widely use DPX files to store digital scans of photochemical films. The DPX format was developed for this purpose and it can store information about the density of the photochemical emulsion. To be represented as an image, the density information needs to be interpreted into a luminance value. As Rice elaborates,

> somebody is having to decide "this much density means black," that the bulb at the expected brightness won't be able to pass through at all. But it's not necessarily a concept that makes sense for digital media, because if you have an MPEG-2 file, it's not like you can put an equivalent of a brighter bulb in and then the light changes.[46]

Contrarily, in FFV1, there is currently no way to express the notion of printing density. Rice continues: "DPX's vocabulary is based on very analog film semantics. The meanings are very local, and it wouldn't be so natural to have them exist in digital formats. But if we're using FFV1 for film data too, we'd have to be able to figure out how to bring these local meanings in and store them."[47]

It is analytically productive to follow through with the notion of semantics brought up by Rice because it allows us to view formats as statements within a discursive network.[48] The "semantics" of any format can be understood as the "laws of existence of statements" constituted through relationships of power across various institutional mechanisms,[49] to which we will turn in a moment. These "laws," articulated through techniques like standards documents but also through habitual practice, delimit what "can be said" in a format. Even digital formats are thus inscribed with and constrained by the logic of older media like photochemical film. Such historical relationships between formats that subtend both analog and digital media are also reflected in archives' deliberate working protocols. Many traditionally film-oriented archives, like the EYE Film Museum Netherlands, carried the materiality of the film strip over to new workflows. They like to work with formats like DPX and TIFF, which are sequences of thousands of individual image frames. Film archives like EYE see little benefit to working with self-contained video files like FFV1. On the other hand, both Sound and Vision and CUNY TV Library and Archives work with video files, in which the meaningful unit of operation is not an image frame but the file itself, which follows from their proximity to broadcasting institutions.

There are benefits and disadvantages to both image sequences and video files. A single image frame is much easier to restore from backup when it is damaged compared to a potentially large video file. But a lengthy sequence of hundreds of thousands of images has considerably more system overhead on a storage medium than a single file. For this reason, some archives, such as the Austrian Mediathek, opt for a compromise and produce segmented video files with a short duration of one minute.[50]

Even though traditional film authorities like EYE prefer image sequences, these may not always be optimal for smaller archives. Kieran O'Leary, the former Access and Digital Collections Developer at IFI Irish Film Archive, has documented on his blog the hardware-related limitations at his institution in 2016: The archive's film scanner had a 12-bit sensor but could only output images in 10-bit DPX or 16-bit TIFF formats. Out of both lossless and uncompressed formats, the former throws away two very valuable bits of dynamic range, whereas the latter adds four unnecessary bits to each scanned frame, a storage burden consisting of 25 percent redundant data:

> It is a bit frustrating knowing that a quarter of each LTO-6 tape will be filled with zeroes, even more so when each tape has two backups! Lossless compression offsets this redundancy, making it more financially viable to preserve the full output of the sensor for the long term. Even if your scans do not have this kind of extra padding, the storage savings may be significant.[51]

Even a digital zero weighs something. And when your machine generates billions of them, that weight adds up. It is in situations like these that the materiality of digital data becomes palpable and the very contemporary lossless digital video standards resonate with a historical aversion to wastefulness that has accompanied many format standardization processes across media history. One final example from Dave Rice:

> Sometimes people are using film scanners to record four-channel video or red, green, blue and infrared, but then they end up storing it in formats that can't support an infrared plane, so they just hide the infrared in an alpha channel. Which is well-defined and supported, but it ends up being a hack, because you're storing data in an alpha channel that's not transparency. So, it's a consideration for designing a format: how do you define the intent on a plane-by-plane basis? If you're doing a film scan using more complex filtering systems—so you have red, green, blue and other shades—how would you be able to store all the planes independently?[52]

Very few image and video formats can intelligibly store information about the infrared layer of color, but such information might be valuable for archivists working with photochemical film.[53] The lack of robust support for such use cases can become a point of loss if the data is preserved improperly or with the semantic provisions of a format not designated for it. Moreover, this problem is decidedly not confined to digitized analog material. Every born-digital asset is also filtered, sampled, quantized, and compressed at many points during its creation and preservation. As Rice relates, a current FFV1 draft seeks to add support for Bayer array encoding to accurately represent the position of sensor elements in a video camera or film scanner. An optical sensor contains different sensing elements for red, green, and blue light, arranged in a tiny lattice according to a pattern known as a Bayer array. Common video codecs in use today fuse information from neighboring sensor elements into a single pixel with red, green, and blue values. But, strictly speaking, this is not what a sensor is "seeing," since each color is scanned at a slightly different position in space and there are usually more green sensors than red and blue ones to mimic the human eye's higher sensitivity to green light. Therefore, preserving detailed information about the structure of the sensor lattice might end up being salient to preservation.

Such technical subtleties might seem insignificant or negligible to us today, but with increasing resolutions and technical possibilities, good archival practice would dictate that these differences matter, or at least that they may matter to preservation in the future. Does the loss of exact information regarding Bayer arrays constitute a "loss"? This example neatly illustrates the material intricacies of a concept as seemingly straightforward as "losslessness." Depending on how we choose to define loss, even lossless formats frequently lose information in a multitude of ways, for example on the technical level of bit depth, dynamic range, or simply wrong description of what the data represents. But this case also shows how archival understandings of losslessness are necessarily impermanent. When, in the future, multi-wavelength scanning or detailed Bayer array information becomes common enough among many archives with sufficient institutional power, not storing this data might retrospectively start to appear as a loss. This redefinition of loss might then incentivize all archives to strive to adapt their scanning and preservation workflows. It is this hindsight of technological history that makes some formats—or rather, some practices—appear lossy. It therefore seems productive not to tie losslessness down to its narrow meaning in electrical engineering and not to reduce it to a single variable, such as the bitstream. Without the necessary infrastructural and social work of developing format standards that can meaningfully contain new information, such as Bayer array data, new processes will remain "lossy," as inter-institutional differences in the storage and description of data will lead to creative hacks and workarounds like those described by Rice: Archivists in one place might decide to store Bayer data in a different way from their peers elsewhere. These differences will then necessitate new consensus-formation and standardization efforts which may eventually result in yet other lossless standards. The understanding of losslessness continuously keeps slipping away as new imaging, filtering, scanning, and compression technology is introduced.

The issue with video codec standards, as with most other sociotechnical objects, is that their real-world implementations are astonishingly messy. It is one thing to have a standards document or format specification and it is another altogether to implement that standard in a real, operational process. While the standard aspires to define a perfectly unambiguous ideal, it is implemented by people who make mistakes and machines that mess up. Its concrete embodiment in physical objects or material procedures is unlikely to be perfect. Peter Bubestinger-Steindl has tested a range of lossless video encoders, including those for FFV1 and JPEG 2000 in 2017.

Much like Rice's examples with glitchy LTO tape machines, he found numerous issues, including lack of support or incorrect implementations of certain features and files that could not be verified as lossless.[54]

Both Rice's and Bubestinger-Steindl's experiences of inscrutable failure teach us an important lesson about the functioning of all standards and underscore the difficulties obstructing archivists' work: People who work in the preservation of video cannot fully trust nor rely on the tools or software they use. Altogether, the above examples demonstrate that a reformatting in the digital domain is anything but a question of simply copying some numbers but entails very much the same—perhaps even exponentially more complex—material considerations as an analog format change.

LOSS AS A SOCIAL RELATION

Apart from interferences on the level of machines and incompatibilities between formats and specific preservation systems, contemporary archival practice heavily relies on cataloging tools and techniques like metadata, which also follow their own formatting standards.[55] Any sizable collection will further necessitate database and media asset management systems and depend on information-organizational techniques like file segmenting and folder structures. It will rely on scripting languages, browsers, and graphical user interfaces that make the use of various command-line tools easier; on software libraries, ingest machines, tape recording machine suppliers, and digitization management software.[56] Archivists will further make use of conformance checkers and will be guided by various guidelines and flowcharts, risk assessment, and planning tools. Their work can be affected by licenses, recommendations, and standards, written and unspoken conventions, and "best practices" such as those developed by actors like the Internet Engineering Task Force, UNESCO, IASA, and others.

Despite their practice and knowledge having so much to do with the non-textual, moving image archivists rely on a vast, growing archive of written text. The specialist knowledge and techniques that archivists access and share—for example, how to program a script that will perform some type of verification on a collection of files, how to implement a specific feature or deal with a temperamental machine—is often generated and circulated in semi-formal and informal contexts, mailing lists, white papers, online discussion boards, presentations, blog posts, and face-to-face conversations.[57] The examples given by our interview partners demonstrate how methods are often improvisatory, but *ad hoc* workarounds for spontaneous problems can turn into accepted solutions and protocols simply under the force of repetition. All the above elements and actors follow their own—sometimes complementary, sometimes conflicting—economic, ethical, aesthetic, and political rationales.[58]

For losslessness to function as intended, numerous interacting layers require careful balancing, beginning with the minute technical characteristics of a signal and reaching all the way to long-term institutional policies. Each single node in this "socio-technical system,"[59] or what we might call *archival assemblage*, can potentially be in conflict with any other. One important finding we might derive from this is that "losslessness" is not really a technological property of any individual video standard or codec. To echo film archivist Roger Smither writing in 1987, "[t]here is [. . .] a tendency to assume that the introduction of new technology in some way automatically results in an improvement in services and circumstances. This is simply not confirmed by experience; the improvement must be worked for."[60] Losslessness takes a lot of labor. While it is convenient to think of compression algorithms as clearly definable digital "things" with well-defined properties from which certain effects can be derived,[61] in the context of present-day media, the functioning of video standards really rather depends on a vast network of ancillary elements, pre-existing machines and workflows, knowledge exchange among various communities of practice,[62] and forensic techniques of verification. The real locus of new video format standards is not in the codec but diffused across these multiple overlapping surfaces. All of them need to be oriented in just the right direction and executed in the correct order before they can begin to generate the desired outcomes.

Workarounds and other "ill-structured solutions," as Susan Leigh Star might have called them,[63] have been a source of fascination for sociologists of information technology and media scholars for decades.[64] In many respects, the way archivists collaboratively sort things out in the archival assemblage is no different from how similar informal organizational processes and structures had been described in academic literature forty years ago, back when the sociological study of malfunctioning copying machines, early artificial intelligence systems, and biologists was *en vogue*. But some of the conditions under which this work is performed now have also changed radically. The ways of participating in particular communities of practice, the means of distributing knowledge, and the shape of social bonds are now so manifold and fragmented that establishing anyone's belonging to a particular group or organization is even more complex than it already was in the 1980s. The new possibilities of engaging in (or withdrawing from) sociotechnical and standardization work—from social media to videoconferencing, from wikis to GitHub—demand continuous attention to the fine shifts in how people and machines encounter each other.

Jonathan Sterne maintains that obsolescence is a social relation as much as it is a relation of technology.[65] We could transpose this to loss, too. The effects of lossless video standards emerge not from some hidden algorithmic principle but from the interactions of many competing rationalities. In the notion of losslessness resonates that very old promise of immutability which "new media" keep making, yet which never seems to fully materialize. There is a critical difference between what a compression codec or format promises to do, and the material costs required to avail of this promise in practice. Shared understandings of loss are both dependent on and transformative of the social relationships that practitioners in the preservation field negotiate in order to perform their work. To say this implies that losslessness is ultimately more of an ontological rumor than an attainable reality. It is best understood as an imagined line that perpetually recedes into the distance. It is reactivated in specific historical forms, much like related notions of fidelity or authenticity, and, as the sound scholar Melle Kromhout adds, their close theological agnates "purity," "eternity" and "immortality."[66]

This does not mean that losslessness does not cause tangible changes in the archive or that it does not have concrete and useful effects on archival work. Contemporary ideals of losslessness—and the many failures that result from their necessarily imperfect implementation—dynamically respond to as well as guide technological developments and new format standardizations. On a material level, lossless video standards do not truly institute a media-technological form of immutability but rather simply induce new forms of loss. Despite archivists' best efforts, losses and failure are never quite banished from the archival assemblage. As Dave Rice emphatically told us, "[t]here's so much failure! All the time. I feel like often we think we are adding a lot of things into our workflow that feel a bit paranoid, but then just kind of end up finding that the paranoia is justified because they end up detecting many problems."

The Changing Face of Archival Labor

The institutional notion of "loss" as something that can be verified on the level of the bitstream rather than as something within human sensory reach represents a significant epistemic break in media preservation. Older preservation strategies at the cusp of the digital transition—for example, the Performance Model of the National Archives of Australia from 2002—still spoke of an "essence of a record": "those characteristics that are essential to the meaning of a document's message."[67] The Performance Model adopted the conceptual separation of an abstracted "message" from its material carrier which became prevalent in digital preservation after the 1990s,[68] but explicitly excluded the bit order of the data as non-essential. While this older approach relied on the judgment of a human archivist to establish what the essence of a record was, mathematically lossless formats have shifted the locus of the debate. The bitstream became reified as the essence of a digital record. This conceptual reorientation has also led to changes in archival practice and the introduction of new techniques, skills, and tasks. Exemplary among them is checksumming, an algorithmic technique that now performs a task traditionally assigned to human archivists.

THE CHECKSUM

A checksum algorithm takes a long input and produces from it a short value—the "sum." The checksum is a machine-generated and machine-readable alphanumeric code that functions as a type of signature: If even a single bit within a file changes, that file will produce a completely different checksum, so comparing the checksums of two digital files is a quick way to verify whether their bitstream is identical.[69] Provenance—the proof of an unbroken chain of custody from donor to archives and a central tenet of archival and museum work—can be difficult to prove when working with digital files, which are easily copied and reformatted. Checksums are one method favored in many digital preservation sectors as a way of documenting provenance and authenticating whether a particular file has sustained any "losses" (that is, unintended changes to its bitstream) over time.

Checksums are useful tools but lack the kind of granularity that large video files require. For example, if a single bit in a 20-gigabyte video file flips, that file will fail its checksum validation. Frustratingly, this failure usually does not tell the preservationist exactly where in the video file the damage resides, meaning that repairing the file may be impossible and a completely new file must be created—a time-consuming and, depending on the quality of the original source, sometimes impossible task. In the words of Ashley Blewer and Dave Rice, "[t]he larger the file is, the less effective a checksum mismatch is at clarifying the extent and location of the error."[70] Newer video compression formats like FFV1 alleviate this issue by incorporating checksumming techniques directly into the encoder itself.

FFV1's version 1.3 has the inbuilt capability to verify the integrity of image data internally at the level of a single frame slice. The codec divides each image frame down into slices, or chunks of the lines that make up a video image. The Matroska container format can additionally embed multiple checksums for a digital object, meaning that combining the granularity of FFV1's slice-level encoding with the Matroska format can give great precision with respect to where checksum failures occur in a file.[71] Kieran O'Leary from the Irish Film Archive gives a concrete example on his blog. When performing a backup of files digitized from a Betacam SP tape, an automated checksum comparison reported that an FFV1 file had not been copied correctly. Out of curiosity, O'Leary found the error by using the internal checksums that FFV1 provides. But upon viewing the corrupted frame, he found that he could not see any visible damage (Fig. 1.3). This was because FFV1, like many compression codecs, also performs automatic error correction: The codec can mask damage by using picture data from previous frames, essentially acting as its own film restorer. In O'Leary's words, "[t]his was so successful that I could not see where the errors were, without using specialist tools like QCTools."[72]

Such granular checksums can, in theory, allow the preservationist to precisely target the repair of a media file. In our interview, Ben Turkus of the US Library of Congress told us this "internal fixity" of the FFV1 bitstream makes it very attractive to him, and in chapter 3, we discuss in detail how the FFV1 compression format supports other functionalities that preservation professionals find useful for their work. But here, it is worth pausing to reflect on O'Leary's report and his quest to locate invisible damage to a digital file.

The use of lossless compression and algorithmic tools like checksumming transforms how archivists engage with archival material. Previously, the relationship between the archivist and the film in her care was centered around sensory techniques: it is by judiciously mobilizing their skills of looking and careful observation, but also with the trained "literacy" of touching and smelling a film or tape or listening to the acoustic traces of a machine's operations that archivists would monitor decay and damage. The recent reconfiguration of what loss means blurs the circumstances that enable an archivist to immediately distinguish between what is or is not decay. Much like in the case of Dave Rice's mysteriously failing LTO machines, in O'Leary's example, a novel form of loss has crept into audiovisual material without being accessible to sensory inspection. Verifying the presence or absence of such losses requires algorithmic intermediaries such as checksumming tools and software that can analyze temporal differences between video frames. Algorithms now constitute crucial infrastructures of the moving image, and they are increasingly vital to its preservation, too. Losslessness is primarily understood, enacted, and verified algorithmically. Therefore, examining how algorithmic format

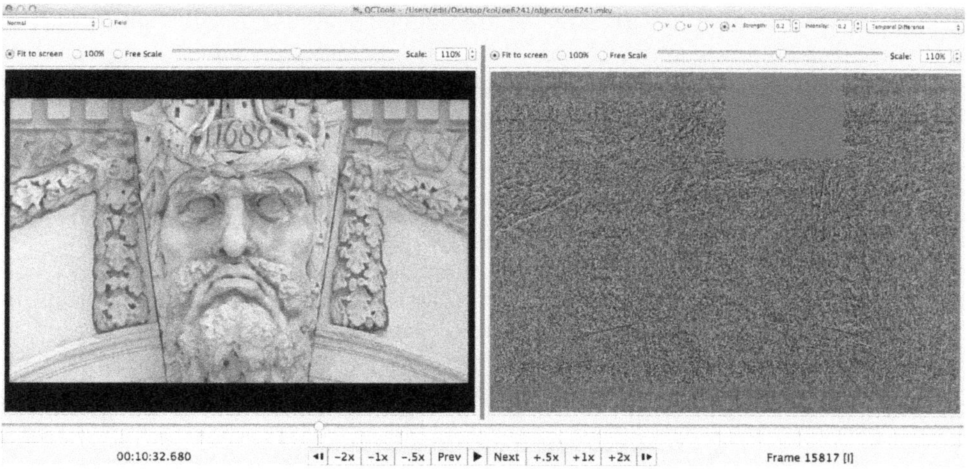

FIG. 1.3. Left: A seemingly intact video frame as visible during playback. Right: a temporal difference filter reveals a solid gray block in the top portion of the image, indicating damaged data. (Screenshots of QCTools interface by Kieran O'Leary.)

standards function, both technically and as tools used by archivists in everyday practice, is a prerequisite in understanding the present and future of the moving image.

The use of electronic assistive tools for fault detection and integrity verification is nothing particularly new, and certainly not unique to digital environments. Devices like vectorscopes that can visualize signal irregularities invisible to the human eye have long been used in quality control in analog moving image production, broadcast, and preservation. What is new, however, is how lossless compression, error correction, and algorithmic tools like checksums transform the way archivists can engage with archival material. O'Leary investigated and documented the damage out of curiosity and to learn more about the error, but there is, in principle, no need to look at the video file at all—an archivist could remedy the damage simply by repeating the copying procedure as instructed by the fixity check.

This can be contrasted with the checks performed at EYE Film Museum Netherlands when it ingests new digital material. As described to us by Ivo Noorlander, digital access staff at EYE, the color space of digital films that production companies deliver to the archive tends to be wrong "all the time."[73] But in EYE's case, the key verification tools used to test for such errors are an archivist's eyes and aesthetic sensibility: The archivists check the incoming files manually and adjust the color space when it appears wrong. Another compelling case in the context of still images is raised by archivist Laura Capell, who documented the digitization of a decaying photograph collection:

> In the most deteriorated negatives, embrittlement and shrinkage resulted in extensive channeling, which made it difficult to see the image. Other negatives had blue streaks across the image. The archivist [. . .] was prepared to discard them, based upon the Libraries' unwritten policy of discarding items that are damaged beyond the point of usability. However, because of the subject matter of the negatives [i.e., historically significant events], the archivist hesitated. [. . .] Materials documenting university history are of particular importance as the Libraries prepares for the University of Southern Mississippi's upcoming centennial celebration. For this reason, the archivist looked for ways to preserve the negatives instead of discarding them.[74]

This is precisely the type of historically sensitive and context-dependent decision-making that human archivists are able to perform in order to intervene in the preservation process. Despite the "unusable" state of the records, the archivist recognized the limitations of this notion of usability, opting instead to disregard policy and preserve the material even with its historical damage. Fully automated and algorithmic preservation systems are unable to account for such edge cases. When the fixity of archival objects is upheld algorithmically, an archivist will no longer be necessary—or capable—to verify changes to the records. Instead, the archivist's role becomes to monitor and maintain the software-based procedures.

This is a recent development in film preservation that demands attention from media studies and archival studies. The process did not begin with lossless formats but is exacerbated by them. Historically, it emerges from the interplay of other computational techniques like error-detection, motion estimation algorithms, or the early electronic film dirt concealers of the 1980s.[75] Recent economic pressures—among them, the broadcast and film industries' increased hunger for archival footage—are also intensifying the process.

A pessimistic view of this transformation might warn that the disappearance of sensory skills "elevates machine vision and calculations, while denigrating educated human perception as reliable verifier of truth claims and quality."[76] One could draw a historical comparison with the work of film projectionists. Until the late 1920s, film projection was considered a form of creative artistic work, and projectionists were vital contributors to the performance and aesthetic realization of a film.[77] With the mechanization of film projectors and standardization of 35mm frame rates around 1930, a rapid devaluation of projectionists' labor occurred. From then on, projecting a film meant the menial activity of monitoring a machine, rather than the creative task of shaping the dynamism of a film's performance. As increasingly large portions of archival and restoration work are performed automatically, is it not conceivable that a similar danger awaits preservationists, destabilizing their sensory perception as a source of knowledge and thereby deprofessionalizing their role as experts?

A more hopeful view might observe that the labor of moving image preservation is simply diversifying, has never been stable to begin with, and that exciting new forms of film preservation are emerging that do not necessarily involve ever seeing a film. It is telling that neither of the preservation experts we interviewed saw new algorithmic techniques and formats as a threat. Rather, while they certainly agree that a striking shift is taking place toward preservation as a form of software engineering, this is in itself seen as a new field of specialization in which archivists can be the experts. Some of our interviewees voiced misgivings about the increasing necessity to understand code and the comprehensive amounts of damage that mistakes in designing software workflows can cause to a collection. But overall, the new face of film and video preservation as software engineering and data management is seen as a welcome development that could raise the profile of the archiving community as a whole.

The new type of film preservation described by O'Leary, Rice, and others requires expertise and knowledge of code, a familiarity with Python scripts and FFmpeg and hex editors, and an understanding of what checksums and network protocols are and how they can and cannot be used. These new skills and more traditional film preservation duties are not mutually exclusive, and often required of the same person. Yet as an increasingly important, autonomous mode of archival practice, they have not yet been well studied. Even if acceptance of algorithmic tools is high among the experts we interviewed, it does have profound ramifications. Questions like what the new techniques will specifically mean for smaller archives, which might tend to rent equipment or outsource services externally, remain to be answered.[78]

To our question about what skills an archivist needs to have in order to work with lossless video formats, some archivists—such as Ashley Blewer—responded that "if we keep building tools, and tools keep supporting the formats, archivists won't need to have as many advanced skills [. . .]. Ideally, the skills are the same as working with any other video format."[79] Yet, more often, what we heard in interviews and observed in the institutional structure of the archives we have visited was that archivists trained in photochemical or magnetic preservation often have trouble working with digital material, and vice-versa. Lossless digital formats require a hugely different set of skills, often to the point where these sometimes operate, basically, as two secluded archiving communities. Justina Akporherhe, principal film archivist at the Nigerian Film Corporation, recounts that

> [T]wo years ago, there was another department that was merged with the archive. And that is the laboratory department. The laboratory department was merged with the archive because films are no longer being processed on celluloid. [. . .] But because they're not in the system from the beginning, it was very hard for them to actually integrate. It was very hard for us, too, because, [as] I told you, we are just five, and the five of us that were working then, one person was moved to the Film Institute Library. Then the remaining four of us—one is a very junior officer, who is just a clerk, but we have to bring him into the work of archiving, because we just need hands, we need more hands. So, the new set of people that joined the archive, they don't really have the basics. [. . .] So, when it comes to the digital aspect, it's very difficult for them to flow with us. So, that's why . . . most times, the digitization it's just two of us.[80]

Even in digital environments, the material legacies of older media can prove very challenging.

Conclusion

It may seem paradoxical that lossless formats have appeared in preservation practice right around the time that scholars of archives and memory began increasingly acknowledging the value of loss. The literary scholar Anne Whitehead argues that the prevailing focus of contemporary memory studies has, indeed, been the recognition of forgetting.[81] If there is a consonance to be found among humanities scholars studying memory, it is that memory is possible only through mechanisms of loss, forgetting, and failure. The media scholar Wendy Chun has stressed that "*degeneration*—not regeneration—makes memory possible,"[82] and the cultural anthropologist Aleida Assmann similarly argues that "[m]emory, including cultural memory, is always permeated and shot through with forgetting."[83] Many other scholars across a range of fields share this view.[84]

In a 1927 essay, the German philosopher Theodor Adorno speculated that the increasing sonic losslessness of phonograph records—concurrent with the increased industrialization of their production—induced a loss in the quality of musical instruments and in the textured constitution of singers' voices:

> As the recordings become more perfect in terms of plasticity and volume, the subtlety of color and the authenticity of vocal sound declines as if the singer were being distanced more and more. [. . .] The moment one attempts to improve these early technologies through an emphasis on concrete fidelity, the exactness one has ascribed to them is exposed as an illusion by the very technology itself.[85]

Today, we would ascribe this passage to Adorno's one-sided rejection of the culture industry and perhaps shun it as technological determinism. But the core of Adorno's thought keeps re-emerging in different permutations within media and archives scholarship. The ascent of digital technology throughout the 1980s, 1990s, and 2000s was accompanied by a strange mixture of anticipation and unease. As archivists looked toward the promise of preservation without loss and decay, media scholars like Mary Ann Doane warned that digital technology would annihilate the imprint of time inherent to all analog media, and thus wipe out a sense of history.[86]

Ultimately, the lossless transfer of data that digital media are nominally capable of has turned out to be possible only under very specific circumstances. "The digital" is too broad of a brush to theorize archives and memory. Many digital *formats* are lossy and forgetful—even quite exuberantly so—and even those that are not, are lossless in a small handful of ways only. The notion that digital media could once and for all solve the problems of analog decay does not have as much purchase among archives as it did a decade or two ago. Precisely *because* most digital media are so fragile and fail so often, archives are constantly migrating formats, anxiously monitoring the obsolescence of their storage carriers and integrity of their files, and hiding RAID racks in caves and mountain bunkers.[87]

But even with all these caveats, we cannot completely dismiss Doane's diagnosis because the promise of the digital as "a dream of immateriality, without degradation or loss" continues to inform preservation practice to some degree.[88] It is called upon when archival experts invoke "virtual immortality" at conferences, and it remains in place as a horizon in the concept of losslessness, perpetually just a tiny bit out of reach. The idea of losslessness embedded in format standards like JPEG 2000 and FFV1 is an attempt to enforce order and stability and speaks to an archive's efforts to keep the digital things in its custody in a stable and coherent form. But in practice, objects, including digital ones, are unruly and "fugitive."[89] Losslessness is always impermanent, deferred as soon as it seems achievable—or, to borrow Nanna Bonde Thylstrup's diagnosis of mass digitization, "a vision whose grand rhetoric often works to conceal its messy reality."[90]

This may make moving image preservation seem like a Sisyphean endeavor, always one step behind technological challenges and conceptual slippages. But one could argue that the fact that archivists are able to preserve moving images at all is not *despite* the constant threat of decay and error but *because* of it; because it is in the encounters with failures and mistakes that archivists continuously develop new preservation strategies and modify and improve their ways of doing. Archival practice is forever a bricolage, always in a preliminary state. Many solutions to problems are permanently provisional. Yet this is, we would argue, not a flaw in the system nor a weakness. It is from this instability, from the hybridity and messiness of doing, from the interferences and errors, from the complicated politics archivists must navigate that new things emerge.

Studying the formats and standards that live in an archive means studying the past, present, and future of the moving image, all at once. This chapter has provided a systematic analysis of losslessness, an emerging notion in audiovisual preservation whose full conceptual, historical, and pragmatic ramifications still remain uncharted. The adoption of lossless formats marks a tectonic shift in the status of cultural heritage and memory. With the introduction of lossless video standards, "loss" no longer exclusively refers to qualities that humans can perceive, define, evaluate, and account for, but rather is the result of a numerical comparison performed by algorithms.[91] We have argued that "losslessness" needs to be understood not in technological and absolute terms, but as a symbolic horizon that structures a dynamic assemblage of practices, institutions, and technologies.

Based on interviews with preservation experts and leaning on previous critical scholarship on digital media and archives, this chapter reaffirms the stubbornly lossy materiality of digital formats in archival contexts. Previous studies on the materiality of compression and digital video have been overwhelmingly interested in their lossy forms. This is only fair, as lossy compression underlies the defining media formats of our time. It is at the heart of all the photos we take with our phones, every streaming platform we subscribe to, and every digital music service we listen to. Lossless compression, on the other hand, has received very little attention. But it deserves to be critically interrogated with equal force, not least because lossless technologies are used to store much more permanent if less visible forms of knowledge: forensic and criminological files, the financial records of banks, genomic data, and, increasingly, archived film and video. When the immutability of such archival records (and with it the social legitimacy of archives) is grounded in algorithmic processes, "the forensic imagination may impose forms of knowing that reproduce particular power relations."[92] Although lossy and lossless formats belong to opposing registers in the class hierarchy of images, the power of lossless compression is just as central to memory, not least because it aims to make memory immortal.

Lossless video standards not only complicate the conceptual foundations of moving image preservation, but they also significantly change what it means to be a film or video archivist. After having described the material functioning of archival video formats and their transformational effects on the ways in which audiovisual archivists perform their work, in the next chapters, let us look at the intricate politics of how and why standards are developed and adopted by archives.

2

Standards with a Capital S

The Making and Meaning of JPEG 2000 and MXF

Standards act as boundary objects: technologies that "inhabit several intersecting social worlds."[1] Boundary objects can have different roles for different users and, perhaps more importantly, also have different meanings for different people. Video format standards operate at the boundaries between media producers, distributors, exhibitors, and—ultimately and emphatically at the tail of the sequence—archivists. For decades, the media preservation community has had to rely on hand-me-down technologies—machines and software—from the broadcast and cinema industries to carry out its digitization and preservation work. The archives and library communities have rarely collaborated with standards bodies like SMPTE, EBU, or MPEG (the go-to authorities in all matters related to film and video technology standardization) to develop these tools.

Over the past decade, this has been changing. Instead of working with the traditional arbiters, a significant portion of the preservation community has been looking in a new direction. The Internet Engineering Task Force, or IETF, is an increasingly important—though unusual—partner in the audiovisual standardization effort. The IETF is the organization responsible for standardizing Internet infrastructure. The decision to standardize some archival video formats at the IETF raised many eyebrows among audiovisual engineers. Why should an organization responsible for *Internet* technology meddle in the standardization of archival *video* formats? Understanding format standards as boundary objects will help us see that the answer ultimately involves big ethical questions of openness, access, and democratic participation in the standardization process.

The following two chapters explore this gradual increase in archivist and librarian participation in the development of video standards.

Serving as opposites are two pairs of video formats that represent different approaches to standardization: the formal standards JPEG 2000 and MXF, and the informal open-source formats FFV1 and Matroska. This chapter looks at the former, documenting the relationship between the traditional media industry and the preservation community, and the motivations that guide large audiovisual archives like the US Library of Congress to adopt formal standards influenced by television broadcasters and film companies.

A Complicated Relationship: Hollywood, the Broadcast Industry, and Memory Institutions

Audiovisual preservation and digitization are steeped in highly technical standards. Video playback and capture involve technological artifacts that are often unique and irreplaceable. Once they are no longer produced, they are nearly impossible to reverse-engineer and reproduce. For most audiovisual playback and digitization devices, it would require an entire factory to create replacement parts. The means of production of film and video technologies are so complex and costly that most archival institutions—which do not tend to generate much, if any, revenue—rarely have as much influence over this industry as broadcast and film production companies. This lack of influence by libraries and archives was a factor that swayed Indiana University to choose open-source standards, as Andrew DaPuzzo, from the digitization company Memnon who worked with the university on a large audiovisual preservation project, pointed out to us. The same desire for more autonomy motivated many other archives to follow suit.

The science and technology scholar Langdon Winner notes that technology adoption often requires social and political arrangements.[2] SMPTE, the Society of Motion Picture and Television Engineers, can be seen as a socio-political and economic arrangement between media producers, hardware manufacturers, and software developers. This standards body is made up of professionals and businesses from all of these endeavors: Some of its most prominent members include the likes of Disney, Sony, Google, Microsoft, Apple, Arri, and dozens of other media technology manufacturers, streaming platforms, conglomerates like Warner Media, large broadcasting corporations like the British BBC and Japanese NHK, alliances like the European Broadcasting Union, and others. Although anyone can become a member and archivists certainly do participate in SMPTE's activities, one would be hard-pressed to find an archive among its corporate members. Because audiovisual technology is continually evolving in terms of visual and aural characteristics like resolution, dynamic range, color rendition, 3D, virtual reality, and so on, developing tools for production is usually more profitable than developing tools for the capture and playback of obsolete formats. This is one of the central conundrums that archives face: The formats they rely on were developed by an industry looking for novelty and innovation, not preservation of old materials.

From the NTSC and PAL standards that describe various aspects of analog video signals to the digital encoding and format standards that are the focus of this book, standards are boundary objects that mediate between the worlds of memory institutions, media production, and media dissemination. But audiovisual preservation is a niche market, which means many institutions have had to build workflows and budgets around procuring, repairing, and employing existing, cast-off devices and technology. Josh Yocum, one of the media preservation specialists we have interviewed, states in a guidance document he created for Northwestern University, that libraries and archives often find themselves "piggybacking on the standards used by the recording/broadcast/movie industry."[3] Part of this piggybacking owes to the nature of the video signal and the preservation mandates of many institutions. Because the "ancillary" data that travels along with the actual content of a program—closed-captioning or show titles and summaries, for example—may have value to the historical record, archives often find themselves on a parallel mission with broadcasters who also want to create tools for clearly capturing such ancillary information.[4]

Another reason for the archive's dependence on industry is a widespread belief among preservationists that the production realm has already anticipated most potential problems with a particular technology and that such technology is thus fully formed and trustworthy. As we will see, this is not necessarily the case. As our interview partners Nick Krabbenhoeft (Digital Preservation Manager at the New York Public Library) and Jim Bird (President of Post Production Services at the digitization and restoration company Scenesavers) both pointed out, large institutions like the Library of Congress can have a strong influence on the choices of smaller shops that do not have audiovisual expertise on staff. In these cases, the reputation and size of these entities give them a certain power over the decisions made by others who do not have similar influence.

In yet other cases, archives decide to turn the media digitization work over to vendors rather than perform it in-house—a situation that can lead to what we have described earlier as path dependence. The archive becomes dependent on the techniques, limitations, and format decisions of the vendor—occasionally with accumulating technical issues and errors, as we have seen in the previous chapter—and the same vendors are used over and over by more institutions.

Archives and libraries are thus enmeshed in a system that moves technology from the world of media production into that of cultural memory and heritage via a network of vendors and suppliers, and finally into the waste stream where its lives end. The story of the media industry's influence on archival formats can be told especially well through two of its main protagonists: the container format MXF and the compression standard JPEG 2000.

New Standards for a New Era

THE MATERIAL EXCHANGE FORMAT

American television broadcasters developed the MXF format in the 1990s during the early stages of the transition from analog production to digital file-based workflows. In the 1990s, the American broadcast industry was beginning to move from a magnetic tape-based paradigm to digital asset management infrastructures.[5] Engineers realized that in a digital environment, poor asset management practices make it much easier to lose "assets" (as the many components that make up a film or TV broadcast are often called). For this reason, MXF was developed to "package" digital assets with metadata to make them findable and self-describing. The Library of Congress, in the mid-2000s, determined MXF to be a suitable container for their digital objects as well, and for similar reasons.[6] George Blood, a widely used American audiovisual preservation and digitization vendor, explains why MXF is appealing to audiovisual archives: In 2011, he wrote that "the MXF wrapper was developed with broad industry and user support . . . as it is more widely adopted, vendor support and interoperability proven, and experience grows, it is an excellent choice in the long term."[7]

According to Bruce Devlin, one of the developers of MXF, "[t]he Material eXchange Format (MXF) is an open file format, targeted at the interchange of audio-visual material with associated data and metadata."[8] MXF was designed to encourage the interoperability of files within production environments, ensuring ease of use between file servers, audiovisual recording equipment, and editing tools.[9] This means that MXF is an example of a coordinative standard which, as Raymund Werle says, is "needed to facilitate the interoperation of components of networks and the provision of services in technically heterogeneous environments."[10]

WHAT IS A CONTAINER FORMAT?

MXF is a container format. Let us use the analogy of a letter in an envelope (a favored example in the literature[11]) to briefly explain what this means.

Letters come in different but usually standardized paper sizes. We send letters in envelopes, which also mostly have standardized dimensions. Often, the envelope will be smaller than the sheet of paper, so to make your letter fit, you must make it smaller. You could cut the paper to fit the envelope—the equivalent of lossy video compression—but this is an inelegant solution, especially for an archive, which would prefer to keep the whole letter in its original state and structure. A better alternative is folding the paper, which will allow recipients to unfold and read it as it was originally written. There are different conventions for folding letters to make them fit an envelope: a business letter is often folded in half or in thirds, and a personal one might be folded twice into quarters. In this example, the letter is a video file, and a compression codec like FFV1 or JPEG 2000 is the folding method that makes it smaller. Different codecs will "fold" the video file in different, more or less efficient ways to save space.

The container file format is like the envelope. The benefit of using an envelope is that you can enclose and transport other things—a photograph, for instance—along with your letter in one conveniently self-contained package. A container format does the same. It bundles a video file together with one or multiple audio streams (different language versions, different sound mixes, or a director's commentary, perhaps), subtitles and closed captions, timecode information, images (such as a film poster), and other data describing this content into a single, compact digital file (Fig 2.1).

Envelopes come in different sizes and can have different properties—some are sturdier, some are self-sealing, and so on. Not every envelope is suitable for all types of documents. In much the same way, digital container formats can also have different properties, and not every type of video encoding can be stored in every container format. The process of standardizing a container format like Matroska or MXF is thus the equivalent of inventing a novel type of envelope with unique properties, convincing enough people that your envelope is a great fit for many types of letters, and then collectively agreeing upon a set of instructions that would allow anyone to mass manufacture envelopes that exactly follow those specifications. Getting the standard adopted, then, is analogous to convincing stationery shops, printer manufacturers, manufacturers of automatic letter-sealing and letter-opening machines, and other technology to support your type of envelope.

One of the chief purposes of the MXF format is to ensure that a media file is self-describing: it carries bundled within itself the metadata a human user or an operating system may need to properly understand or decode the file's information. This is important, because metadata (such as information about the color space, interlacing, timecode, audio channel arrangement, the language of closed captions, and suchlike) is sometimes

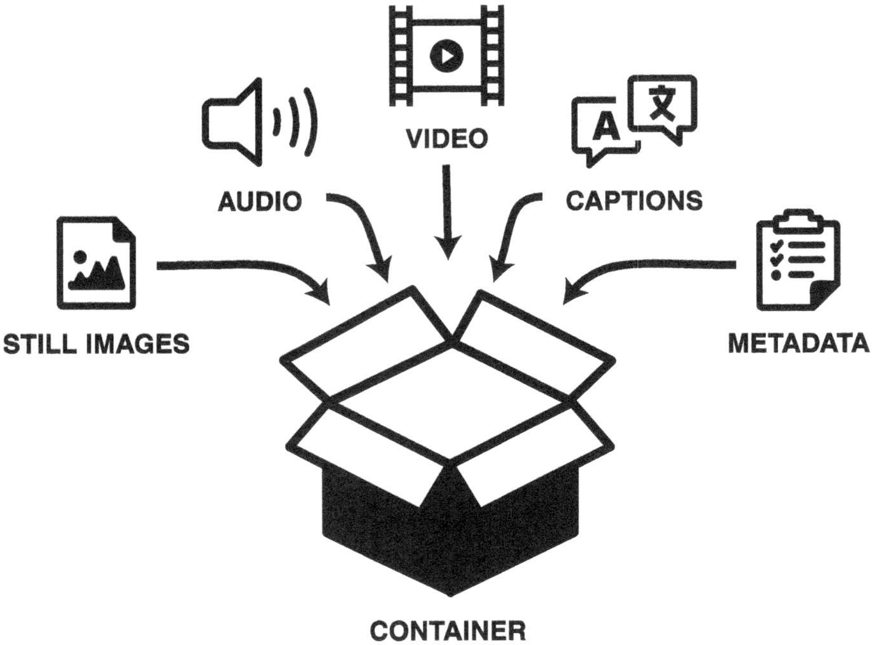

FIG. 2.1. A visual representation of a digital media container format. (Diagram by the authors.)

stored in so-called sidecar files separately from the content, and those files run the risk of being separated from the main video tracks. Imagine an envelope with a bunch of photographs in it. If you describe the photographs on separate snippets of paper, it is likely someone might lose them in the future, and it will be impossible to identify what the photos show or when they were taken. It is better to note this information down on the photographs directly or on the envelope—or both. Libraries and archives share with the broadcast industry the need to ensure that digital objects can move about freely in IT environments without the risk of separating important descriptive metadata from the contents.

MXF: A VERY BRIEF HISTORY

MXF was born out of a forum within the Moving Picture Experts Group called Pro-MPEG, and, at least initially, designed primarily for television and radio broadcasters and not for the world of cinema. In the late 1990s, Pro-MPEG was testing various digital formats for interoperability within production environments. At the time, it was becoming clear that production environments were soon going to be running on fully realized digital infrastructures. The needs of broadcasters would not be served by any existing, "off-the-shelf" format because existing audiovisual file formats were not able to handle the needs of high-quality and high-throughput production systems. For example, the AVI (Audio Video Interleaved) video container format, while ubiquitous, was poor at containing video timecode information in a predictable, standardized way.[12]

Concurrently, the media tech company Avid, the producers of the nonlinear editing suite of the same name which is used throughout Hollywood and television broadcasts, was working on a tool they were calling AAF, the Advanced Authoring Format, which they were designing for very similar ends.[13] Representatives from Avid joined the Pro-MPEG group's discussions and told the group that what they were intending to do—to create a file format that would be useful to a broad array of producers, software manufacturers, broadcasters, and hardware producers—was unlikely to move along in any efficient way. After months of analysis and negotiation between members, the Pro-MPEG Forum realized that there was little likelihood of getting production hardware and software designers and manufacturers to agree enough to develop a new *encoding* standard, that is, a common way of compressing video and sound files. However, the forum came to realize that there could be sufficient agreement and appropriate buy-in for a *container* format for the interchange of files between systems and software.[14] Thus, the Pro-MPEG and the AAF group began to work together to draw up the initial specifications for MXF in 1999. However, the group ultimately decided to standardize the MXF format with the SMPTE rather than at MPEG because of its more formalized, due process structure.[15] MXF was ratified as a SMPTE standard in 2004—a major achievement for the broadcasting industry: "for the first time, we have a standard way of bundling together all components of a programme, making it simple to interchange rich content."[16]

Some of the major contributors to the development of MXF were the BBC, Grass Valley, Panasonic, Sony, Pinnacle, and Omneon. Panasonic was particularly invested in MXF because of their interest in a file container that could encapsulate the various digital objects produced when their cameras captured video essences and associated data to their new P2 solid-state capture card technology. This investment by Panasonic in MXF is an example of a very common phenomenon in standards-making that we will encounter in chapter 4, too: that of for-profit industries trying to influence standardization in order to gain or maintain market share for their products.[17] As this outline illustrates, the power of the SMPTE name with respect to trust in particular standards is substantial in the domains of radio, television, and film production. We will return to the cachet of SMPTE standards later, but first, a brief description of JPEG 2000 is in order.

THE JPEG 2000 ENCODER

JPEG 2000 is a still image encoding that takes its name from the Joint Photographic Experts Group, the working group within the International Standards Organization that developed and wrote the standard. JPEG 2000 is a successor to the original, ubiquitous JPEG image compression algorithm and the intention of this newer standard was to "minimize the number of bits required to represent the original image samples without any loss of information."[18]

One of the principal powers of JPEG 2000 (and what differentiates it from its predecessor) is that images encoded in the newer format can be decoded at any level of image quality up to and including its highest. This makes JPEG 2000 highly useful as a still image encoding in digital libraries because it allows for image scans that can scale in quality in proportion to a user's Internet connection bandwidth. The original JPEG standardization work began in the 1980s, with a draft international standard released in 1991. Of the standards under discussion in this book, only the original JPEG existed before the World Wide Web. JPEG files have a very high image-quality-to-file-size ratio: The format can reduce the size of an image file tremendously while resulting in only moderate losses in quality, which manifest as the familiar blocky compression artifacts.[19] These compressive talents made JPEG very amenable to dissemination on early, dial-up Web connections, helping the format achieve its ubiquity.

In the late 1990s, photochemical image capture was still the norm for many Americans. Increasing access to the Internet meant more people were scanning their film photographs and uploading them online, which caused the number of JPEG files in the world to balloon. Thus, the JPEG group began considering a new, more scalable version of the standard. In 1997, the JPEG working group published a call for new compression technologies to be submitted for their fitness evaluation inclusion in a new image format. The ISO standardized JPEG 2000 as ISO/IEC 15444-1 in December of 2000.[20] Some of the major players in the testing and standardization of JPEG 2000 were Kodak, Booz-Allen & Hamilton, Mitsubishi, Ericsson, Texas Instruments, Motorola, the Sarnoff Corporation, and Hewlett-Packard.[21] While there was some representation from nonprofit entities like the Norwegian University of Science and Technology or the University of Maryland, many of the most influential actors in the standard's development were for-profit businesses.

Over the course of the early 2000s, a SMPTE working group called DC28, the Technology Committee on Digital Cinema, began investigating technologies to chaperone Hollywood's burgeoning transition from photochemical to digital cinema distribution. SMPTE chose JPEG 2000 as the compression method for moving image sequences in digital cinema, using the technology to encode each frame of video as a still image. JPEG 2000's winning characteristics were its ability to scale images and its lossless compression scheme,[22] although cinema projection uses a lossy form of the codec. The term "Motion JPEG 2000" (MJ2K) is often used in this context, which is simply a sequence of individually encoded still images, even though it was never officially standardized as such.[23]

SMPTE also chose MXF as the container for these digital cinema materials, which helped this format combination achieve major representation not only in television broadcast but in the film industry, too. This SMPTE Digital Cinema Initiative included significant involvement by seven major Hollywood studios.[24] The studios' close involvement led to MXF and JPEG 2000 becoming ensconced in Hollywood and in the American television and radio broadcast industries, as well as in the archives and libraries that preserve their materials.

MXF, JPEG 2000, and the Library of Congress

MXF and JPEG 2000 moved into cultural memory institutions in the early twentieth century when the Library of Congress began investigating them as a possible target container/encoding format pair for their analog video digitization work (Table 2.1). In a post on the email list for the Association of Moving Image Archivists, video engineer and preservation specialist Jim Lindner mentions that

> JPEG 2000 was never positioned as a forever format. It was always discussed as a transition between analog then (around 2002) and the next one (whatever that will be). It was one link in a migration chain that extends outward in time. The really important part at that time was that it had mathematically lossless compression. Compression was a big deal at that time and considering the size of files and the cost of storage at that time consensus in the broadcast industry and elsewhere was to use lossy compression for archival storage of AV content. It was a real struggle to convince people to not use lossy compressed files.[25]

The Library of Congress, because of its enormous collection of moving image materials, began working with Lindner in the early 2000s to explore using JPEG 2000 for its high-quality digitization.[26] Lindner

Table 2.1. How MXF and JPEG 2000 moved from their niches into libraries and archives

<table>
<tr><th>MXF Container Format</th><th>JPEG 2000 Image Compressor / Encoder</th></tr>
<tr><td>Developed by the SMPTE television broadcast community in the 1990s</td><td rowspan="2">Created by the ISO JPEG as a high-quality, scalable online still image encoding in late 1990s</td></tr>
<tr><td>Used in television production and archives in late 1990s/2000s</td></tr>
<tr><td colspan="2">Became part of the Digital Cinema Initiative in the 2000s</td></tr>
<tr><td colspan="2">SAMMA started using it in early 2000s</td></tr>
<tr><td colspan="2">LOC began working on FADGI MXF/JPEG 2000 specification in 2009</td></tr>
<tr><td colspan="2">FADGI specification comes under AMWA auspices in 2012 and is named AS-07</td></tr>
<tr><td colspan="2">FADGI version of AS-07 is published by FADGI in 2017</td></tr>
<tr><td colspan="2">Specification is renamed Registered Disclosure Document (RDD) 48 and published by SMPTE in 2018</td></tr>
</table>

developed the first JPEG 2000 video encoder and integrated it into a device called SAMMA, or "System for the Automated Migration of Media Assets." SAMMA is a customizable stack of cassette video playback decks that can be set to automatically digitize several tapes (Figs. 2.2 and 2.3). It is a "set it and forget it" system that is especially useful for high-throughput digitization in large media collections. The SAMMA system, while initially developed for the Library of Congress, became attractive as a time-saving digitization tool for other institutions like the Smithsonian Archives and the Peabody Collection. The SAMMA technology successfully migrated outside the United States, as well.

With SAMMA came JPEG 2000 and MXF. The device generates high-quality JPEG 2000 files with the option to wrap those files in the MXF container. As this piece of hardware moved into memory institutions, so did the format standards. Hardware support for digital standards is critical to their growth and adoption. Sterne points this out in his discussion of the development and spread of MP3 players like the Zune in the late 1990s and early 2000s. As Sterne notes, these playback devices allowed the MP3 to become an on-the-go music format in the mold of the cassette tape and the Sony Walkman decades earlier.[27] Without reliable hardware to encode the analog video signal into a JPEG 2000 bitstream and wrap it into an MXF container, the standards would not have been adopted by preservation institutions, both in the United States and elsewhere, such as at the Cinematheque Royal in Belgium, the Institut National de l'Audiovisuel in France or the City of Lausanne Archives.[28] As it turns out, the digital format needs of archives and libraries are, often enough, similar to broadcasters and film producers: storage space is usually inadequate for the wealth of

FIG. 2.2. A single SAMMA digitization deck at the Library of Congress's National Audio-Visual Conservation Center, Culpeper, Virginia, USA. (Photograph by Jimi Jones.)

FIG. 2.3. Multiple SAMMA digitization machines at the Library of Congress's National Audio-Visual Conservation Center, Culpeper, Virginia, USA. (Photograph by Jimi Jones.)

digital content being created or acquired, and media objects need to be self-describing for transport and ease of retrieval.

However, not all institutions found MXF/JPEG 2000 files easy to create and decode. Some archivists were suspicious of relying on a unique piece of hardware for fear of vendor lock-in, as we will discuss in more detail in a moment. Other institutions like Indiana University, upon testing the files generated by the SAMMA device, found those files had interoperability problems with other software.[29] SAMMA, while useful for some institutions, was not a cost-effective or usable solution for all.

The LoC's preference for the MXF/JPEG 2000 combination owes to more than just their use of SAMMA. The library houses the United States Copyright Office and one of the stipulations of the assignment of copyright to a motion picture is that a copy be deposited into the library's collections. Because Hollywood production companies were investing heavily in the development of a digital cinema standard using MXF and JPEG 2000, it was immediately clear to the library that these kinds of files would begin flooding into its collections once digital cinema came to maturity.[30] The large volumes of digital film and video material entering the LOC from the American television and film industries include more than just moving picture information. One of the reasons the library chose the MXF container is because of its ability to contain closed captioning information along with video and audio essences.

While the term "standard" might seem to imply a fully stabilized, uniform technology, the documents relating to the MXF and JPEG 2000 formats betray a labyrinthine network of specifications for various implementations or "flavors" of the formats. A wide array of different vendor implementations of the standards exists, and these implementations lack conformance with one another.[31] The .mxf extension is not a guarantee that a file will be readable on an MXF-compliant machine. The software engineer Pedro Ferreira relates that

> MXF is quite a complex standard and many of its concepts were quite new for the industry. The sheer complexity of the undertaking inevitably led to some lack of precision in the language of the standard, some corner cases that had not been foreseen, some workflows that had not been properly addressed, some implementations that were not flawless, etc.[32]

As sketched in the introduction, this is a relatively common occurrence with format standards, but it can cause chaos and confusion in archival institutions with respect to the specifications to which their files and their encoding and decoding software and hardware must conform for proper production or playback. And that is why the Library of Congress sprang into action.

A VERY STANDARD-LIKE NONSTANDARD: AS-07

At the time of this writing, the Library of Congress, together with other members of the US Federal Agencies Digital Guidelines Initiative (FADGI) has spent over a decade developing a specification for using MXF and JPEG 2000 in their video digitization work. This document, once called AS-AP (Application Specification for Archiving and Preservation), is a model case for industry-influenced standards development. Since at least 2016, the specification has been called AS-07 by the Library of Congress.

A chief purpose of the AS-07 specification is to ensure the uniformity of outputs from the Library of Congress's video digitization workflows (Figs. 2.4 and 2.5), maintained by specifying a subset of the MXF file format for the long-term archiving and preservation of moving images. Another important goal is to describe the library's video files with enough precision so that vendors can develop hardware and software that can create and decode those files. AS-07 is called an "application specification," an in-between object that Geoffrey Bowker and Susan Leigh Star might have called a "residual category" without the full binding power of a standard,[33] which nonetheless acts as a standard for interchange between technologies. Much of the Library of Congress's work on AS-07 happened in face-to-face meetings with engineers at the Packard Campus of the National Audio-Visual Conservation Center and in meetings of the US FADGI Audio-Visual Working Group.

The wide array of conflicting and idiosyncratic implementations of MXF was an important reason the Library of Congress decided to draw up a document that would clearly lay out the library's preferred implementation of MXF and JPEG 2000 as a kind of "recipe" for use in their preservation work. FADGI, of which the library is a part, began drafting their specification for MXF and JPEG 2000 in 2009. From 2012 to 2015 FADGI used the Advanced Media Workflow Association, or AMWA, to help them develop the specification document. AMWA is an industry body comprised of media companies and their suppliers that developed the AAF format and, together with the Pro-MPEG Forum, the initial specifications for MXF. The library chose AMWA as the incubator for this new document because of the organization's ties with broadcasting and its track record of drawing up similar documentation for the BBC. The AS-07 development team included private vendors and developers like George Blood Audio and Video and the Belgian audiovisual tool company EVS-OpenCube.[34]

James Snyder, former video engineer at the Library of Congress, was a strong proponent of developing AS-07 because the standard would allow for ease of interoperability between compliant files and tools.[35] Furthermore, the AS-07 document would provide a recipe for hardware and software developers to ensure that their tools could be used by (and marketed to) the Library. In the words of Carl Fleischhauer, one of the principal authors of the specification, "AS-07 has been defined broadly, as a file format for the preservation and long-term archiving of audio-visual content, ranging from theatrical motion picture content to broadcast and non-broadcast video content to sound recordings as separate entities."[36]

As Kate Murray, head of the FADGI Audio-Visual Working Group, reminded us throughout our interview, FADGI is not a standards body and they do not see the AS-07 specification as a true standard. It is, as she notes, more of a "recipe" for generating video files with very clearly defined characteristics. She also told us that there is no "one format to rule them all" with respect to preservation-quality digital video, and that this means the AS-07 specification can have an educational component

FIG. 2.4. Video digitization facility at the Library of Congress's National Audio-Visual Conservation Center. (Photograph by Jimi Jones.)

FIG. 2.5. Digitization decks for various video formats, including VHS and U-Matic at the Library of Congress's National Audio-Visual Conservation Center. (Photograph by Jimi Jones.)

for other institutions trying to make decisions about their own digitization work.[37] While Murray described AS-07 to us as, for the moment, not being a true standard—insofar as it has not been created by a recognized standards body—the language of the document adopts parlance peculiar to standards. It behaves like a standard in that it rigorously describes the elements that video files must contain to conform to the specification. The document describes levels of constraints that range from "gentle" to "moderate" to "strong" to put conformance to the specification on a continuum.[38] For example, if a user generated a video file with a gentle level of constraint to the AS-07 specification with respect to the number of lines of resolution of the video image, the file would *likely* be decodable with any MXF-compliant software. However, if the user conformed an aspect of a file to a strong level of AS-07 constraint—in this case the precise number of lines of resolution called for in the document—the user could be *assured* that this file would, if only in this particular aspect, precisely mirror a similar file produced by the Library of Congress.

So while there is no power to enforce the FADGI specification outside the Library of Congress, any vendor, donor, or other user who wanted to create tools to create or decode files for use with the Library of Congress, or any user who wanted to make files in line with the library's workflows, would probably be safest adhering to the strongest possible level of constraints within AS-07. In this way, the AS-07 document functions precisely like a coordinative standard—by focusing on interoperability and compatibility. As James Snyder told us, the slow, methodical standards-making and review process of SMPTE is one of the reasons he feels assured that, when a SMPTE standard is finally released to the public for use, it is ready to go "right out of the gate" because, in his opinion, the body has already worked out nearly every bug and potential use case.[39]

Through the course of developing AS-07, the FADGI working group came to see the limitations of MXF and JPEG 2000 in the cultural heritage context. Some technological affordances of MXF work very well in archival institutions—for example, the container has a specified space for the retention of closed captioning data.[40] In analog television, this type of information is often carried in line 21 of the analog NTSC video signal, which is hidden by the physical chassis of a television screen and thus invisible to viewers, but can be decoded and displayed by a compatible device. In audiovisual preservation, this type of ancillary information—beyond simply the visual video signal itself—forms an important piece of the archival puzzle and should thus be preserved.

However, because the MXF developers directed their focus on new, turn-of-the-century digital video capture technologies, certain crucial aspects of legacy analog video were ignored in the initial development of the standard. For example, FADGI discovered that MXF, as of 2011, had not properly defined interlaced JPEG 2000 video—likely because MXF developers did not seriously consider the need to digitize legacy interlaced video in the era of digital production.[41] Interlacing is a video compression method that reduces signal bandwidth by combining image information from two separate frames of video into one through a raster of thin horizontal lines. This process is highly susceptible to errors, and improperly digitizing interlaced video can permanently damage archival materials.[42] The lack of definition for how to handle interlacing complicated FADGI's work and motivated them to petition SMPTE to expand the MXF specification to correct it.[43]

After several years of working with AMWA to develop the AS-07 document, the Library of Congress chose to move the specification into the SMPTE standardization stream. According to Kate Murray, developing AS-07 into a SMPTE standard would give it "more staying power" in the archive because of two reasons: Firstly, the American audiovisual preservation community is comfortable with SMPTE's work, and secondly, standardization by SMPTE would also keep the document in line with the broadcast industry because of its heavy involvement with the standards body.[44] Murray describes the AMWA business model as "not a good fit" with the library and AS-07's move from AMWA to SMPTE jokingly as a "divorce."[45]

The move to SMPTE standardization is also important on technical and political grounds because the library's collections "include significant holdings produced by broadcasters, content with complex added elements like captions and subtitles, multiple timecodes, and other elements."[46] The technical need for capturing information beyond just picture and sound is motivated at least partially by the library's need to be perceived as a good steward of every element of the broadcasting and cinema materials in their charge, in a manner befitting the national library. Murray noted that as they have developed AS-07, they have found issues related to MXF's description and handling of the preservation and digitization of video and FADGI has relayed them to SMPTE for correction.[47] In this way the AS-07 project has been a rare case of the preservation community providing useful feedback to and influencing the work of audiovisual production. As of this writing, AS-07 is still in draft form.[48]

THE CHALLENGES OF THE BLACK BOX

Despite many of its useful features, AS-07, a specification built upon a television broadcast format and employing standards-making groups with high broadcast and Hollywood representation, was met with mixed reactions in the larger audiovisual preservation profession. Part of the reason for AS-07's sluggish adoption is that, while the specification itself is openly available to all on the FADGI website, the tools it is built upon—MXF and JPEG 2000—suffer from a "black box" problem.

The workings of black-boxed technologies are often inscrutable to their users and do not lend themselves to being reverse-engineered, retrofitted, or changed. They also tend to enter markets with a feeling of inevitability—the illusion of being perfectly formed even when, in reality, they may lack closure and not be compatible with all use cases.[49] MXF entered the wider media preservation community via broadcast television, which did not have the same mandates, workflows, or infrastructural limitations as the world of libraries, archives, and museums.[50] With broadcasters, preservation will always inevitably be sidelined in favor of production.[51] JPEG 2000 was initially similarly maladapted for preservation, as Ashley Blewer noted about the "hacked" nature of Open JPEG. Blewer believes that openness in standards can help to obviate problems like what she saw in JPEG 2000:

> It's because they're open that makes them such a good fit for preservation... [The] Open JPEG standard... was really just... a hack on top of reverse engineering JPEG 2000. [W]hat prohibits the ability to do long-term preservation [is] when you can't actually see how they're built.[52]

The preservation community's lack of participation in the development of MXF and JPEG 2000 has led to standards that are not ready for use without significant retrofitting. In the case of MXF, this retrofitting involves time-consuming back-and-forth with SMPTE, which, as Kate Murray noted, is necessary for AS-07 to be useful with respect to digitizing legacy materials. The black-boxed nature of MXF and JPEG 2000 has repercussions in AS-07 and its adoption outside the US Library of Congress.

Key to the difficulty in implementing MXF or JPEG 2000 in preservation contexts is the nature and limits of the technologies themselves. The preservation of interlaced video content is not the only challenge that the AS-07 initiative is facing. As Dave Rice told us, one of the issues in implementing JPEG 2000 for video is that the encoding cannot natively declare its own aspect ratio; it must rely on the file container to record this information. To recall our letter analogy, this is equivalent to having to indicate on the envelope the date on which the letter it contains was written. It is an awkward way of conveying information in an archive, since video files sometimes get reformatted and repackaged into other containers and that can lead to a loss of important information about the aspect ratio. For preservation, it is more secure if such information is stored in the file itself—written down on the letter rather than the envelope, so to speak. Some other formats, such as the FFV1 files we discuss in the next chapter, can declare their own aspect ratio irrespective of container, which can make them a more appealing encoding method.

Many of our interview partners were reticent about the JPEG 2000 format for a variety of other reasons. According to Nick Krabbenhoeft (NY Public Library), part of his problem with JPEG 2000 is that it never got the audiovisual production industry backing it needed to succeed and grow, leading to a video encoding that was never fully mature before it got pressed into service by the Library of Congress.[53] A further challenge to the AS-07 specification itself, according to Lars Gaustad (National Library of Norway), is that it is a highly technical document, developed by video engineers that are very specialized.[54] Brecht DeClercq (Flemish Institute for the Archiving of and Access to Audiovisual Heritage, then VIAA, now Meemoo) agreed that the challenging content of AS-07 is a reason that some archivists choose not to invest in it.[55]

The LoC's work on JPEG 2000 and MXF was initially tied to their investment in the SAMMA video digitization system. The LoC's influence was felt across the Atlantic, too: The National Library of Norway purchased a SAMMA in 2009, largely because of the Library of Congress's use and because, as Gaustad puts it, the belief, held by many institutions of the time, that MXF was going to revolutionize video preservation. However, some institutions like Scenesavers found that the SAMMA hardware, software, and the files they produced exemplify a black-box problem wherein the technology was not amenable to examination or alteration by users. According to Jim Bird of Scenesavers, SAMMA's designers did not anticipate the need to adapt or alter its technology, which led to a system that required continual tinkering to get right, even after it was released to the market:

> Even within the SAMMA product line JPEG 2000 has been sketchy . . . The heart of the SAMMA was this magic JPEG hardware encoder board that we could never figure out where that came from. Jim [Lindner, inventor of the SAMMA system] must have found some manufacturer to build those for him, I don't know who it was, we never could figure out who that was. At the heart of that machine was really a DRC 1500 card [likely a DRC-1500 Streaming Media Capture Card], or I'm not sure of the model, so there was a lot of magic and we were real familiar with everything else about that machine except for that.[56]

Part of the challenge of AS-07 is the relative slow speed with which it has been developed. This slow development process has allowed other tools—notably Matroska and FFV1—to find purchase with institutions that had at one time considered MXF and JPEG 2000. The broad flexibility of MXF may be part of the problem with its adoption. Its developers could not foresee all potential implementations of the container, which led to many "homebrew" variants of it, as well as many SMPTE extensions of the standard.[57] The lack of a standard SMPTE extension developed specifically for preservation has led to many institutions experiencing incompatibility problems with encoders and decoders.[58] Audiovisual technology vendors have, in recent years, become more withholding of the inner workings of their tools, which further complicates the compatibility of files across software tools.[59] Once again, the black box becomes problematic for MXF/JPEG 2000 users.

Path dependence also remains an ever-present specter. Upon finding a software and hardware combination that works, archives can be loath to experiment with other technologies.[60] This institutional inertia can leave an archive or library in a precarious position if the vendor changes its technology or goes out of business completely. The Library of Congress, because of its reliance on SAMMA and its close ties to broadcast and digital cinema standards, runs the risk of lock-in as well, as James Snyder noted.

Concerns about vendor lock-in and path dependence make AS-07 unpalatable for some institutions. As Hermann Lewetz says, the Library of Congress made its decision to push developers to make solutions because the library has the money. According to Lewetz, the danger is that other, small institutions will look to the library and try to emulate their solution, which would be too expensive for them. This belief is shared by Kieran O'Leary at the Irish Film Archives, who found AS-07 to not be useful for low-budget shops that do not have strong ties to vendors. Indiana University tested AS-07 but also decided against it because, in comparison to other options, AS-07 presented "greater technical complexity and a lack of available tools for this option. It was felt that JPEG 2000/MXF presented greater risk for us over time."[61] As recently as 2016, the FADGI AS-07 test files showed errors which made them problematic for testing. But the LOC's associations with the media industry as well as the sunk costs—the resources invested into SAMMA and AS-07, as well as the thousands of files the library has already created—force it along its path. There may also be a bit of face-saving, too. As Jim Schupert of Scenesavers noted, if the library moved away from MXF/JPEG 2000, it would require telling Congress and the audiovisual preservation community that this expensive journey has not resulted in quite the zeitgeist that the library thought might come of this venture. It should be noted that, as Lars Gaustad said in our interview, there is still a place for AS-07 and MXF/JPEG 2000 in larger repositories with access to resources to support the standards, and the Library of Norway may still adopt something like AS-07 if the specification becomes more broadly accepted. However, it does not appear likely that smaller institutions with more austere constraints will be able to easily implement AS-07 or similar format standard combinations.

Conclusion

The story of audiovisual media heritage and preservation is at least partially the story of the influence that the American television and cinema industries can exert on nonprofit institutions like libraries and archives. This influence can also be felt at the level of the standards that govern digitization and digital preservation.

Industry-influenced standards have been a critical part of audiovisual preservation since its inception. The media industry can leverage far more economic and technological weight over the design of new moving image and sound capture technologies than most libraries and archives. As James Snyder pointed out, even the Library of Congress has very minimal influence over the standards employed by television and cinema producers in the United States. In the Library of Congress, because it is the nation's copyright submission office, MXF and JPEG 2000 easily move between the production and archives boundary. The flow of materials over this boundary is part of what informs Snyder's view of MXF and JPEG 2000 as being standards appropriate for theatrical and television dissemination and archival preservation.

Over the past few years, the Library of Congress has heavily promoted its specification for the use of MXF and JPEG 2000 for their video digitization. Other archival institutions around the world have closely watched the library's work on this specification to see if it can be adapted to their own needs. The implementation and spread of technologies like standards can be very heavily influenced by the amount and kinds of resources that different actors can bring to bear upon their promotion. For example, the Library of Congress has a larger voice in the audiovisual preservation profession than the City of Vancouver Archives does.

But the format standards used in audiovisual production and preservation are rarely developed in anticipation of future needs because those needs are rarely entirely clear. This owes largely to the difficulty in foreseeing how audiovisual technologies will change over time. While it is true that MXF was developed because of the then-impending shift from carrier-based conservation to carrier-independent digital infrastructures, the standard was, at least initially, a reflection of what the standard developers could prognosticate within a limited context of late-90s broadcast media. The resulting standard was, in part, a reflection of the lack of a specified way of handling interlacing in video. The SMPTE MXF working group of the time did not anticipate a context within which the capture of old interlaced video would be a priority.

There is a way of interpreting this oversight as a form of politics: In this case, the broadcast industry's prioritization of new technologies over the old is encoded into the design of the format standard. This oversight represents, in Langdon Winner's words, how "design or arrangement of a device or system could provide a convenient means of establishing patterns of power and authority."[62] The JPEG 2000 image compressor and MXF container were inscribed with a vision of the user and the user's environment that ran counter to the needs and mandates of archives. Because end users are often unaware of the efforts and politics that go into standards-making, it is usually only in cases like this, when a standard "breaks" and its limitations are made manifest, that these politics can become clear.

As in any technological enterprise, audiovisual preservation relies on standards largely because of the need for interoperability between the digital objects being created and the various tools and environments with which they will interact. James Snyder told us that he views standards as being guarantees of interoperability. This view also informs many discussions in the world of cultural memory preservation. Uniformity among objects in a digital repository can be very useful as it minimizes the infrastructure needed to create and play them back. It is not simply the fact of a technology being standardized that ensures its utility for a memory institution, however. There are many competing format standards that require different software and, in some cases, different environments to implement. Some preservation specialists like Nick Krabbenhoeft insist on a format parsimony in their work: implementing the smallest possible array of standards for encoding and packaging their digital media objects.

Furthermore, as Jim Bird noted, the wide array of potentially idiosyncratic implementations of standards like MXF and JPEG 2000 can make SMPTE's backing of the Library of Congress's video digitization guidance documents mean very little. If, ultimately, there is a multitude of MXF and JPEG 2000 file mutations and encoding and decoding software in the wild, how does an institution benefit from strict conformance to AS-07? One of the questions that must be asked when studying a standard and its utility to the world is how the standards hold users accountable for creating objects that do not conform to those standards.

Once again, the broad array of MXF and JPEG 2000 files in the world troubles the enforceability of the standard. And, as discussed in the Introduction, Apple's decision in late 2018 to declare JPEG 2000 a "legacy format" that will no longer be supported in future iterations of its Final Cut Pro could be evidence that JPEG 2000 could soon come to face an impermeable boundary around non-linear editing software.[63] Encodings and containers change very quickly—often more quickly than archives and libraries can anticipate. Another challenge to enforceability lies within the MXF standard itself. As Lars Gaustad pointed out in our interview, there is no singular SMPTE MXF standard. Instead, there are dozens of standards documents consisting of extensions, revisions, and implementations of the core, original document. This has allowed for highly elastic interpretations and diluted the standard's power.

The CELLAR group, to which we turn in the next chapter, is currently facing a similar headache: how to create a standard that is parsimonious enough to *be* a standard but is still a large enough tent to include the past fifteen years' worth of interpretations of Matroska from all over the world. As Bowker and Star put it, this is the fundamental challenge of any standard—balancing the tension between broad interoperability and clearly delineated interpretive flexibility.[64] As we will find out, CELLAR is attempting to neutralize that tension by including many different types of players in the standards-making: for-profit businesses, archivists, librarians, open-source coders, and various kinds of users of Matroska, FLAC, and FFV1.

The development of these format standards confirms the received wisdom within science and technology studies that the best technology does not always win out, and that there may, in fact, not be a *best* technology—that a standard is the result of an evolution of compromises and negotiations leading to a specification that had the virtue of being the minimum set of technological affordances that the standards body members could agree upon.[65] At the same time, Matroska and FFV1 are the most unlikely candidates for a standard for reasons we will discover shortly.

FLEXIBLE STANDARDS, PERMEABLE AND IMPERMEABLE BOUNDARIES

MXF, JPEG 2000, FFV1, and Matroska began their lives outside of moving image preservation. They have all, over the past two decades, moved into use in many kinds of libraries and archives. The developers of MXF designed it as a standard specifically for the American television broadcast industry and, while it is still widely used there, the Library of Congress's

AS-07 specification has helped to expose it to the worlds of libraries and archives. JPEG 2000 has seen a similar move. It was standardized by the International Standards Organization (ISO) for use in online still images. Although it is nowhere near as widespread as its older sibling, the omnipresent JPEG format, the film industry has adapted it for use in moving image dissemination as part of digital cinema. The DCP format, the digital counterpart to what 35mm film used to be, compresses images with the JPEG 2000 algorithm. In the case of the Library of Congress, its use of MXF and JPEG 2000 is largely driven by powerful and high-profile donors to their collections such as Twentieth Century Fox and NBC, who constrain its ability to choose which standards to implement—hence its decade-long investment in the AS-07 specification.

MXF and JPEG 2000 were both developed in that time-honored, classical standards-making mode by SMPTE and the ISO. But this modus operandi, coupled with the high cost of the standards documentation, caused several of our interviewees to be less drawn to these format standards than other alternatives. SMPTE, because its makeup consists largely of Hollywood and broadcast specialists and technicians, strongly directed the development of MXF toward the needs of those industries, as evidenced by the technical and material issues related to interlaced video that came to light during the early AS-07 work.

Before the Library of Congress's work on AS-07, libraries and archives had little input in the development of MXF or JPEG 2000 as a video encoding. As Kate Murray told us, AS-07 may become a SMPTE standard over time, but as of this writing, archival institutions (outside of federal agencies like the Library of Congress and the National Archives and Records Administration) have still had little input in MXF's evolution. Some of our interviewees noted problems making AS-07-compliant files on their own software and hardware infrastructures, exacerbating the challenges that MXF and JPEG 2000 are having with respect to finding a home in many archives.

Interpretive flexibility—the ability for a technology to have different meanings for different users—is critical to making that technology function as a boundary object. Some of the four standards analyzed in this book operate better as boundary objects than others. Two other formats are central to the archival endeavor: Matroska and FFV1. In the next chapter, these formats will usher us into a world, an ethos, and a community quite different from what we have seen. FFV1, developed as an experimental lark by the early developers of an open-source video app, has seen increasing adoption among European and American libraries, archives, and broadcasters. Matroska spent its formative years as a format used almost exclusively by online video pirates but has matured into a robust container technology embraced not only by preservation professionals but also by content producers like Sony as part of their consumer video playback technologies. There is considerable interpretive flexibility in the cases of Matroska and FFV1. The next chapter will show how they have become boundary objects largely because of the looseness and openness of their specifications. Much of the standards-making work plays out over email lists, GitHub threads, and synchronous, open online working groups. This is a provocative counterpoint to the traditional, closed-door standards-making meetings that may require travel and membership costs. However, as we shall see, not everyone is sold on this way of making a format standard just yet.

3

Wild Formats

The History and Standardization of FFV1 and Matroska

When the video compression format FFV1 was publicly presented to the preservation community, the biggest grievance aired against its archival use was that it had not been ratified by any formal standardization body. Many of our interviewees cited the lack of SMPTE or MPEG involvement as cause for concern about adopting it. Strangely enough, for others, this was a selling feature. With this contradiction as a point of departure, this chapter documents the history of the video formats FFV1 and Matroska and their gradual ascendance from vagrant origins in informal practices of political critique and piracy into archives and, eventually, "proper" standardization.

The trajectories of JPEG 2000, MXF, Matroska, and FFV1 as they moved from their niches to more widespread use in cultural memory and audiovisual preservation have been convoluted and, to some of their developers, surprising. This transition was driven not just by the technical affordances of these formats but also by social factors. Some of their technological features *are* especially useful in preservation contexts. But these formats—Matroska and FFV1, lacking an approved standard pedigree—required advocacy by influential users as well as adaptation for the specific needs of libraries and archives to allow for their growing uptake into memory institutions. And this path has not been smooth, either. Resistance to and acceptance of these technologies have been shifting among archivists. This ebb and flow not only teaches us a larger lesson about the complex life cycles of formats and standards; it also shows the archival assemblage as a kaleidoscope of conflicting ideas about the ethics of technology. Matroska and FFV1 are two technologies whose histories are instructive for understanding how formats migrate between formal and informal contexts and turn into standards.

The FFV1 Compression Codec: A Short Overview

FFV1 is rapidly emerging as a dominant codec used by audiovisual archives to losslessly compress video files. The FFV1 algorithm is not exactly new, having lingered around relatively unnoted as part of a popular, widely available, and free video software tool since 2003. It is not exactly new to audiovisual archives, either—a handful of archivists have been using it in their work since 2010. But it has, until recently, been a very obscure format that has yet to become a format *standard*. In August 2021, three older versions of the encoder were published by the Internet Engineering Task Force as an "informational specification." While this is, technically speaking, not a standard, it designates a document "published for the general information of the Internet community" that lends the format some more credence for preservation use.[1] The newest version, FFV1.4, has entered the initial phases of formal standardization at the Internet Engineering Task Force (IETF) in April 2018 and is poised to become a standard in the future.

Not only is FFV1 one of the very few lossless video compression methods usable in an archival context, but its development and inauguration into the standardization procedure is also remarkable. A small comment on an early draft of FFV1's formal specification summarizes the situation succinctly: "This document is from a group of open-source coders, and this is their first IETF experience, so please be extra constructive."[2]

OMNIPRESENT YET INVISIBLE

FFV1 was born outside of any formal institutions or consortia. The video compression codec was created and released in June 2003 by programmer Michael Niedermayer as part of the FFmpeg project. FFmpeg—deriving its name from "fast-forward MPEG" but unrelated to the actual MPEG consortium—is an open-source software library for handling multimedia. FFmpeg is the engine powering nearly all online video. Every major video streaming platform and service, every large social network, and applications like the wildly popular VLC media player use it to compress, decode, and reformat video. The work of the FFmpeg community in reverse-engineering proprietary compression codecs also performs a crucial function in preservation: As Erwin Verbruggen (Netherlands Institute for Sound and Vision) told us, FFmpeg's ability to decode an enormous range of encoders has in recent years partially alleviated some of the earlier fears regarding the long-term readability of obscure and poorly supported video formats.[3]

The programmers working on FFmpeg in 2003 charged themselves with the task of developing a lossless video codec from scratch simply to see if it were possible.[4] Project leader Niedermayer used FFV1 to test whether he could develop an encoding free of the technical constraints of any existing formats. FFV1 was thus being developed a few years after JPEG 2000 began entering media archives. But because of its experimental nature, the codec spent years hibernating in obscurity and without some of the critical attributes needed for work in preservation, including a lack of formal documentation and a lack of support for high-quality 10-bit video.[5]

After lying dormant nestled within FFmpeg for several years, computer scientist and archivist Peter Bubestinger-Steindl tentatively implemented FFV1 at Austria's national archive of sound and video recordings, the Austrian Mediathek, in 2010. Dave Rice from CUNY TV was independently experimenting with the codec around the same time. Niedermayer, Bubestinger-Steindl and Rice, together with Jérôme Martinez, would later all become coauthors of the IETF specifications.

In an interview, Bubestinger-Steindl describes the considerations that led to FFV1's initial adoption in a preservation context.

> After tests with FFV1 we saw that it was the codec with the best performance, good compression ratios and also the most accessible, which was really surprising because we hadn't heard about it from anyone in the professional field before. I first heard of FFV1 in a Russian document from 2004 (updated in 2007) comparing lossless codecs.[6]

The codec's technical affordances were clearly tempting, but the Austrian Mediathek's decision to begin using the then-unknown FFV1 over other, more established alternatives was initially highly polemical. As described in earlier chapters, the archive was met with fierce resistance and even disdain in public forums, confirming the insight made in science and technology studies that "[i]f one does not follow certain standards people may doubt that one is really a particular kind of actor."[7]

But the initial difficulty notwithstanding, Hermann Lewetz—Bubestinger-Steindl's colleague at the Mediathek—reports that many preservation professionals, mostly Europeans but a few Americans as well, contacted him later and inquired about their use of FFV1. Lewetz made the code for his implementation of FFV1 available online alongside test video files, and to his surprise found that there were over a thousand downloads of the files within months of his presentation. As people downloaded and tested the files for themselves, they became much more amenable to the Mediathek's discovery that FFV1 is an effective alternative to JPEG 2000 and uncompressed video even though it does not carry the stamp of standardization that those technologies do.

FFV1 has several characteristics that make it useful in audiovisual preservation. It accepts frame sizes of any resolution and uses either Huffman coding or arithmetic coding for lossless compression,[8] which gives it flexibility. Compared to an uncompressed 10-bit counterpart video file, FFV1 reduces file sizes by nearly two-thirds. That represents an enormous potential for file size savings. It is also relatively fast and efficient: among common lossless video codecs, early tests suggested that it required the least amount of computing power to transcode.[9] FFV1 can also inscribe interlacing information within the encoding, obviating various issues with historical material—discussed in the preceding chapter—that the Federal Agencies Digitization Guidelines Initiative (FADGI) faced with the MXF format.

A TRANSATLANTIC FORMAT WAR? FFV1 VS. JPEG 2000

It was not only the computational appeal and technical affordances, however, that accelerated FFV1's uptake. In many ways, it was, counterintuitively, also its obscurity. Unlike more common encoders and formats like TIFF, whose many mutations, as we have seen in previous chapters, greatly irritate preservation efforts, FFV1 only had a single, simple implementation. It therefore posed few interoperability hazards between multiple components within the preservation workflow.[10]

FFV1's open-source nature meant that missing features—checksumming, support for higher bit depths, interlaced video, and others—could be added with ease and cheaply in later versions, the most recent being version 3 released in August 2013. Version 4 is currently a work in progress at the IETF. Among its features most germane to audiovisual archivists, FFV1 provides multiple checksums for every frame of video, making it possible not only to detect errors and damage in a file but also to identify exactly in which frame it is located. This, as we have seen in chapter 1, can be tremendously useful to digital media archivists. Other considerations such as support for various color spaces and ease of handling also helped its adoption, which has been steadily growing since 2011.[11] However, FFV1 initially remained a very hard sell when it came to actual implementation in libraries and archives.

FFV1 had to compete with a small handful of other methods for losslessly compressing digital video, such as BBC's experimental Dirac codec and their more patrician, older relative JPEG 2000. Driven by format standardization initiatives in the United States documented in the previous chapter, lossless JPEG 2000 had been adopted as a preservation format by the Library of Congress and other federal institutions, as well as by 20th Century Fox and the Library and Archives of Canada. It thus seemed like the prime contender for lossless compression of the digital preservation age. But discourse within the archival community neatly exemplifies the changing tides of format history: Whereas in 2011, JPEG 2000 had been called "[t]he first practical moving image compression standard that doesn't throw away picture content in any way,"[12] only four years later, in the call-for-papers for its annual World Conference, the International Federation of Television Archives was already asking: "Is FFV1 the new JPEG 2000?"[13]

Certainly, FFV1's technical capabilities played a major role in fanning these doubts. But, writing on the differences between FFV1 and lossless JPEG 2000, Carl Fleischhauer, project manager at the Library of Congress Office of Strategic Initiatives, confirms how pragmatic and technological considerations are shot through with an ideological tinge:

> As far as I can tell, the two encodings work equally well: after you decode the compressed bitstream, you get back exactly what you started with, i.e., in both cases, the encoded data is "mathematically lossless." [. . .] At great risk of oversimplification, I wondered if we were hearing from two different (albeit overlapping) communities, each with its own ethos. One community is well represented by national libraries and archives, including the Library of Congress. [. . .] Now: how did this standard-oriented community vote on the lossless encodings? They favored lossless JPEG 2000, a standard from the International Standards Organization and the International Electrotechnical Commission. And the other community? These were specialists—several in Europe—who are strongly drawn to open-source specifications and tools. My sense is that members of this group are eager to embrace formats and tools that 'just work,' and they are less firmly committed to capital-S standards.[14]

Our conversations with archivists indicate that there are, in fact, salient material differences between the two formats in terms of computing power as well as reliability, readability, and compatibility of the compressed output. In our interview partners' experiences, JPEG 2000 is more computationally "expensive," requiring more advanced and costly hardware and generally longer processing times.[15] One of the archives disclosed to us that when testing JPEG 2000 as a migration format on a small cohort of material, they needed an unbearable eighty hours to encode one hour of footage. Many other archivists reported errors and failures in the production of JPEG 2000 files, too, as we have seen previously. Archives and libraries tend to have ongoing resource constraints, so the efficiency of FFV1 is an important part of the format's meaning for those working with it.

As luck would have it, FFV1, too, is now on its way to become a "capital-S" standard and recently also found a spot in the recommendations of influential bodies like the IASA *Guidelines for the Preservation of Video Recordings*.[16] But Fleischhauer's intuition confirms what we have found in conversations with archivists across two continents and what we will elaborate upon in much more detail throughout this chapter: In the past years, as a result of new technological affordances and social reconfigurations in the media and preservation landscape, a clear distinction seems to be forming between audiovisual archives that follow established format standards and those that do not. And the technological affordances of the formats in question are often not the decisive factor at all.

FFV1 evolved from experimental use and resonates strongly with the open-source ethos, whereas JPEG 2000 is used by the film industry and rigorously vetted by standardization bodies like SMPTE and authoritative institutions like the Library of Congress. FFV1's trajectory is an important example of a format traveling on the path from informal practice to formal standard. Both because of this and because of its open-source nature, it is often used in conjunction with another increasingly popular archival format that comes from even more lowly stock: Matroska.

The Matroska Container Format

Matroska, also known by its file extension MKV, is a container format. Much like its more distinguished industry counterpart MXF, Matroska is a vessel that wraps disparate media elements—video, sound, subtitles, still images, metadata—together into a single computer file. In archival contexts, video files compressed with FFV1 are often wrapped in Matroska containers, although this combination is not compulsory. FFV1-encoded video can be stored in several other containers such as AVI (as it was, in fact, at the Austrian Mediathek initially) and the MKV container can accommodate video files encoded with arbitrary codecs. But as with FFV1, MKV possesses several convenient properties for archival work that other container formats do not. Moreover, it also has a fascinating history that starkly contrasts with that of other format standards. Matroska is associated with a milieu quite unlike that of MXF and JPEG 2000: It is the most prevalent format in what Ramon Lobato calls the "shadow economy of cinema"—illegal film torrents, pirated TV shows, and ripped Blu-ray Discs.[17] Search for any current, popular movie on a BitTorrent site and it is with near certainty that what you will find will be Matroska files. In fact, online piracy was also how many archivists we spoke to became familiar with the format.

Matroska has been in existence since 2002 and is based upon EBML, the Extensible Binary Meta Language.[18] It has had wide adoption around the Internet as a format for distributing pirated audiovisual content, likely due to its completely open specification and its agnosticism with respect to the kinds of encodings it can contain. This association with piracy is also almost certainly a reason for its slow adoption within the film and broadcasting industry.[19] The broad use of Matroska in piracy gave the format a kind of ubiquity in the seedier, copyright-violating neighborhoods of the Internet, even as the format's broad implementation led to a wide array of files scattered hither and yon. These idiosyncratic files can lack conformance to shared specifications, meaning that different Matroska files might not reliably play out in similar, predictable ways across playback software types.

Matroska was not specifically designed for piracy, but it *was* designed for use outside the world of established, turn-of-the-twentieth-century media production. Steve Lhomme, a French citizen and now programmer with the VLC video player project, designed the format in 2002 out of a desire to collate and play out clips of the French national political campaigns of the time.[20] Lhomme was collecting video clips to expose the inaccuracies presented by political candidates. But he found the AVI file container to be insufficient for the job of collocating multiple video objects into one container, so he and a partner started working on a format they called TMF, the Tronic Media Format. When his partner left for military service, Lhomme took over the work on his own. The goal of the undertaking was to be able to collect disparate files (and file types) in the same container. Because the "packing" of content within a file reminded Lhomme of the Russian *matryoshka* nesting dolls, he borrowed the word for the use of his new tool: Matroska. On December 6, 2002, Lhomme started generating code for Matroska on the web platform SourceForge, and a new file format was born.

Lhomme himself was a key factor in moving Matroska into piracy: he promoted Matroska on Internet Relay Chat (IRC) sessions and web forums but, as he notes, people were reluctant to use it until they could see the files for themselves. Lhomme uploaded some Matroska files online for public consumption:

> I had a guy on the IRC channel that was there talking to me in private because he was actually one of these pirate people and because they are racing to be the first to release a file or an encoding as soon as the tv show ends, they use the fastest computer you can have and all the

> tools they can think of to be the first to do it. And he created his own Matroska mixer, everybody else was using FFMPEG but he created his own to make sure it's the fastest and he can release this stuff. So he kept asking all kinds of questions about the format and how to either improve it or how to use it better for his use. So that was very interesting. And so this guy gave me an account on FTP . . . where all these pirate people can download [for] free everything that they want. So for a year or two I could download everything I want because I created the format.[21]

Lhomme believes that because he was putting broadcast content into his MKV tests and uploading them online as sample files, he likely tipped off pirates that his new format would be the perfect instrument for their work. This work is something Lhomme is very open about participating in.

A video format developed specifically for purposes of political critique is media-historically exceptional. The media scholar Axel Volmar has emphasized that formats primarily serve to regulate or even enforce cooperation within industries.[22] While their critical industrial functions are undeniable, examples like FFV1 and Matroska show that formats are not always born with an industrial pedigree. Despite its usefulness, Matroska is incompatible—and indeed, undecodable—with professional video processing software. None of the non-linear editing applications commonly used in the media industry, which includes Apple's Final Cut Pro, Avid Technology's Avid, and Blackmagic Design's Da Vinci Resolve, will open Matroska files without additional workarounds. Adobe's suite of video editing and encoding tools had supported the playback (but not the encoding) of Matroska files once upon a time, but support was actually *removed* in 2019 (Fig. 3.1). This is in striking contrast to JPEG 2000 and MXF, which have their own preset in Adobe tools to allow for easy export of such files, and they can, of course, import them, too.

The companies are generally mum about the rationale for the lack of support, citing vague reasons such as low usage,[23] but it is not far-fetched to attribute it to MKV's perceived associations with piracy. Brian Wheeler of Indiana University anticipates that Matroska may take longer to find a

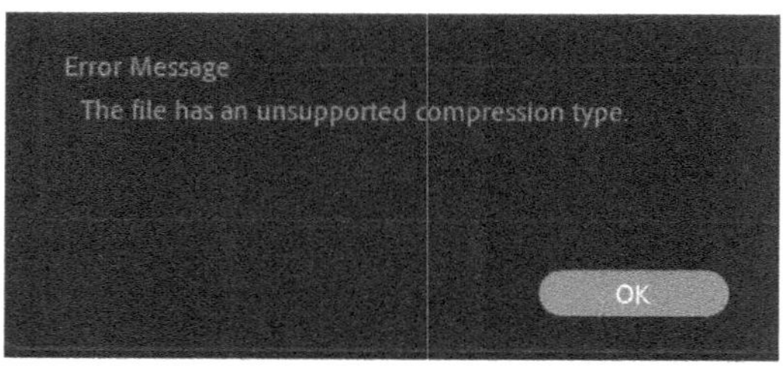

FIG. 3.1. Screenshot of an error message that appears when attempting to import a video file in Matroska format into Adobe's Premiere Pro editing suite (version 24.0).

home in the production world because of this. But the rapid proliferation of MKV files means the format is unlikely to disappear soon. To paraphrase Vincent Mosco, piracy and freeware threaten capitalism's ability to contain them.[24]

WHAT DO AN ARCHIVIST AND A PIRATE HAVE IN COMMON?

Although some large archives are apprehensive about using the Matroska container, there is a curious and unexpected overlap between the work archivists do and practices of illicit file sharing. Much like media archivists, many pirates are also post-production and video engineering experts. In order to circulate films in the high-quality versions now commonly accessible and expected online, they need to have a firm grasp on the filigreed details of signal processing, compression methods, and container formats, and be able to choose hardware that makes it feasible to encode high-resolution, high bit depth and high color gamut films.[25] And while piracy likely slowed Matroska's acceptance in for-profit film and video production, its flexibility has also made it attractive to memory institutions in recent years.

During early tests of FFV1, many archives and libraries argued in favor of using the AVI container because of its simplicity and ubiquity. But that simplicity turned out to be very limiting for audiovisual media because it could not encapsulate all the ancillary data that institutions like the Austrian Mediathek wanted to save. Because Matroska files are self-contained archives that can accommodate content streams of multiple types, they can support some interactive forms of spectatorship, like those associated with the DVD format. If the container includes the appropriate data, viewers can, for example, switch between different language and audio versions of a film, select different camera angles, listen to a director's commentary, switch closed captions and subtitles, or skip between chapters at will. This functionality makes the format highly appealing to online film pirates, but also tremendously useful to audiovisual archivists because they align with mandates common to archival repositories. For example, many American memory institutions strive to make their materials compliant with the dictates of the Americans with Disabilities Act (ADA), either out of choice or legal requirement. As Nick Krabbenhoeft noted, the New York Public Library has found that using the Matroska container for their video materials renders them ADA-compliant more efficiently than other formats because of the ease of embedding subtitles into the files:

> The videos that we were doing are not accessible, and there might be accessibility requirements for those videos in the future, and so being able to run up and get closed captioning done for these materials or any other accessibility requirements that we want and then include that information into the same package was very attractive to me.[26]

Jimi first heard of Matroska while in a meeting of the FADGI at the Library of Congress in 2010. He was a coordinator of FADGI's Audiovisual Working Group and was working on the application specification for the library's use of MXF and JPEG 2000 in their video digitization work. One of the participants asked why the working group was not looking at Matroska as a video container instead of MXF, because of the format's open-source nature. The resistance to this notion came from library video engineers who had concerns about using a technology that was not standardized by a well-known body like the ISO or SMPTE. When Jimi investigated Matroska, he decided that it should be represented in the Library's Sustainability of Digital Formats online resource and created a page in that tool for the MKV format.[27] This is, to our knowledge, the first official acknowledgment of the format by the Library of Congress in a public-facing document.

Over the course of a decade, Matroska's ubiquity in the online world made it difficult for hardware manufacturers to ignore. While Sony initially avoided the format—again, likely due to concerns about facilitating piracy—after a decade of Matroska files circulating the Internet, the company started producing Blu-Ray players that could decode the format. Apple's QuickTime Player can play back Matroska files with the use of a plug-in. In 2011, the City of Vancouver Archives chose to use Matroska in combination with FFV1 compression for its master-quality video digitization. The primary reason for this choice of formats was the Canadian government's requirement that its publicly-funded agencies must use, wherever possible, open and royalty-free technologies.[28] This was one of the first small seams of legitimacy cracking through the pariah format's notoriety.

THE TIDE TURNS IN FAVOR OF FREE AND OPEN STANDARDS

If there was a tipping point in the move to general acceptance for FFV1 and Matroska in the archival community, it was Indiana University embarking upon a massive, campus-wide audiovisual digitization project. In October of 2013, Indiana University President Michael A. McRobbie announced the Media Digitization and Preservation Initiative, or MDPI, which was an effort to digitize all the university's video and audio holdings by its 2020 bicentennial. To perform the immense amount of work needed to achieve this goal, the university partnered with Memnon Archiving Services, a Sony company. According to Mike Casey, MDPI's Director of Technology, by the end of the digitization project, the digital media objects would add up to sixty-five petabytes of data. At the time of our interview with Casey and his coworkers in March of 2018, they had digitized about 5.5 petabytes of moving image and sound content.

When the MDPI project was in its infancy, Casey and his staff and colleagues had to make decisions about target formats and encodings for the monumental digitization effort. They began casting about for technologies that would best serve their work and immediately found that JPEG 2000 and MXF would not suit them because of playback errors in test files and the costs to implement the formats.[29] They also faced the same material constraints—albeit at a larger scale—that many other institutions encounter: They knew that uncompressed video, considered the safest choice for digitization, would take up an unattainable amount of storage space.

As Memnon pressed the Indiana University staff for a decision, they started looking at FFV1 and found that it could losslessly compress their digital video even though, according to Brian Wheeler, "processing time [when encoding to FFV1] was about 25% over uncompressed [video]. It took more CPU time to build derivatives." The processor-intensive nature of the FFV1 encoding process was deemed a small price to pay for compression that can save up to 40 percent of storage space and be encoded more quickly than JPEG 2000 because FFV1 has a simpler compression algorithm. The MDPI staff also found that using the Matroska format as container was preferable to MXF because of its relative simplicity and ease of use.[30] In his 2017 white paper outlining the digitization workflows, Casey states that "[w]e also believe that it is more fruitful, given our specific preservation requirements, to align ourselves with the FFmpeg community rather than with QuickTime developers and Apple."[31]

Casey insists that he intended this statement to be pragmatic and not a rallying cry. But it has, nonetheless, been interpreted by people like Nick Krabbenhoeft as a clear example of the archival community going its own way and lessening its reliance on the production industries. Indiana's choice to use FFV1 and Matroska has had a cascade effect on other large institutions and their video preservation work. In Ashley Blewer's words, it "was huge, because they have a lot of soft power (through good reputation, dedicated money from their Dean for this kind of work, and many big, prestigious collections)."[32] Casey's white paper was, according to both himself and Krabbenhoeft, one of the reasons for the New York Public Library's subsequent investment in FFV1 and Matroska. Ben

Turkus, audiovisual preservation specialist at the library, spoke with a vendor while the library was trying to decide about video formats, and that vendor pushed hard for the Library of Congress's MXF/JPEG 2000 combination. But the openness of Matroska's and FFV1's code and the latter's internal fixity element made that combination much more attractive.

Not everyone in the preservation community agrees that Indiana's direction was sound, though. James Snyder recalled to us some of his objections.

> I've written an answer that I've never sent Mike Casey. [I wrote] an email several years ago when I first saw that white paper. He made some assumptions in there that were incorrect and I was going to ask him about it but I thought, it was probably politically expedient not to do that . . . There are some statements in his white paper that were just flat-out wrong. Like a statement about why they didn't choose MXF, and the reason they didn't choose MXF was flat-out wrong. It means he didn't do his research. Or whoever did his research didn't do their research or didn't understand how it works . . .
>
> We've got the same goal which is long-term survivability of the content, and this is the decision tree that we went through, and you came up with a different answer. Logically we should have come up with the same or very similar answers and yet you came up with a fundamentally different answer. One was based on some incorrect assumptions, but it also looks like your decision tree was different. I'm an engineer, you're a librarian, you're an archivist, a content archivist. We are two sides of the same line. I have to do technology implementations and so I see it from that point of view, how do you enable long-term survivability. When someone is from the other side where they don't know how the technology works and they have to make their best guess, because to them a lot of the terminology that folks like me use is Greek . . . they don't know whether or not they're having smoke blown up their ass or whether or not we know what we're talking about because they don't have the experience to know if what we're talking about is bullshit.[33]

So much of archival practice naturally revolves around technology, but we also see how formats undergird and regulate access to that technology epistemologically: *Knowing* a format means knowing what is right for the archive; knowledge of the format distinguishes "the engineer" from the "mere librarian"—discursively, at least.

However, Rob Mobley of Indiana University made clear in our interview that one of the reasons for IU's choice of FFV1 lay not just in its ability to compress video and thereby save storage space, but because they found the FFV1 encoding to be truly lossless, as opposed to other compression methods.

> After doing a bunch of transcodes back and forth of a sample film we determined that yeah, V210, even though it's sold as being lossless, it really isn't. And once in FFV1 helped us discover that, because it is in fact lossless. [. . .] So we determined it was [the video encoding hardware and software] Black Magic that was breaking the rules but we realized that yeah, you do lose some of the very blacks, very whites, and FFV1 was in fact doing everything very losslessly like we had hoped.[34]

In the last decade, many other big players have leapt into the pool with both feet. In 2013, Britain's national broadcaster and global media giant BBC began to explore the utility of FFV1 wrapped in an MXF container for digitization of some of their video materials. Ian Henderson, an archivist at the BBC, disclosed that they reviewed many encodings and containers for their preservation work but settled on FFV1 and Matroska, largely because formulations of MXF and JPEG 2000 files such as AS-07 were not found to be fully interoperable across a wide array of software encoder and decoder tools. Even large institutions like the BBC can thus find open-source tools useful. In fact, Henderson insisted that FFV1 appeals to him precisely *because* of its open-source nature, as well as the ease with which it can now be generated and transcoded with available tools.

Memnon's Andrew DaPuzzo commented that FFV1 and Matroska handle "where you put stuff flawlessly." Our conversations show that Indiana University's choice of these technologies was motivated not only by the need to save space but also by its experience that the FFV1/Matroska combination is robust and reliable in terms of where diverse types of data are embedded in the container. This experience echoed in several of our other interviews, too. Audiovisual specialists Jim Bird and Jim Schupert at Scenesavers, a Cincinnati-based company specializing in digitization of legacy media, said they see the choice of FFV1 as sound because they had difficulties in implementing other formats and creating files that precisely match the specifications laid out by AS-07:

> But I think the whole SMPTE standard argument doesn't really work in the real world. You can say it's got the stamp but we have had a devil of a time getting that to translate. Wouldn't you think every JPEG 2000 player would play every JPEG 2000 file? If it was a big SMPTE gold standard?[35]

For Scenesavers, having the SMPTE "label" had little meaning if a technology was problematic. Henderson, at the BBC, also chose Matroska instead of MXF because of the former's platform-independence, the wealth of playback solutions for it, its robustness for embedding metadata but also, again, its open-source nature. The *technological* affordances of a format are thus an admittedly critical element of archival decision-making, but they are only meaningful as part of a much larger archival assemblage: How well does the format interface with other formats? How well does it interface with humans? How reliably does it perform in conjunction with other machines and other software?

In making this choice, Henderson found that there was little online information about Matroska and FFV1 in preservation contexts, so he turned to the Austrian Mediathek and the City of Vancouver Archives with questions about their use of these tools. Their support helped him make his choice and the BBC became what was likely the first large-scale repository to adopt FFV1 and Matroska for the preservation and digitization of video.

Since then, many other institutions have also gotten on board and adopted these two formats in concert: the Irish Film Archive, the National Library of Norway, the City University of New York TV, and others. Adoption is growing among broadcasters, too—at least those outside of the United States, who are not as encumbered by excessively litigious copyright regimes or the belief that Hollywood norms represent the needs of all preservation institutions. The preservation expert Reto Kromer noted that there is growing support for Matroska as well as FFV1 among Eastern European broadcasters, such as the Slovenian RTV and others:

> I was very surprised [in 2016]. We had a workshop in Albania with the Albanian Cinema project and I was really surprised that many archives in the former Eastern Bloc of Europe, the former Soviet Bloc, that they use in the broadcast industry a lot Matroska. And if some tools will be available, for instance, to have directly and actively implemented FFV1 in cutting suites or grading suites and programs like Resolve [. . .] then I think they will switch also to FFV1. Because they see the advantages: they have less cost, they have better control [. . .].[36]

This growing adoption of the non-standard formats FFV1 and Matroska gives some credence to both Jim Schupert's and Carl Fleischhauer's assertion that open-source technologies are more readily embraced in Europe than in the United States. Jean-Christophe Kummer agreed with Schupert's sentiment when he observed that Americans are in the SMPTE domain and Europeans are not.[37] At the risk of some generalization, many European institutions exhibit a pragmatism that somewhat shields them from the dependence on "capital-S standards" of their counterparts in the United States.

However, this does not mean that European institutions are impervious to path dependence or unconstrained by their own sociotechnical legacies.

For example, Erwin Verbruggen from the Netherlands Institute for Sound and Vision pointed out to us that, for a large archive like Sound and Vision, "there's constraints to what our setup can do. Our storage management system at the moment, if we would throw FFV1 at it, it would not know what to do with it, because none of those vendors are geared towards that format."

And thus, while it is generally not too difficult to store FFV1-encoded video, in this instance, due to the path this archive follows and the way its practices emerge out of its cultural position, proximity to broadcasters, and specific preservation microculture, the introduction of a lossless standard like FFV1 is currently neither feasible nor considered necessary. A transition would likely require extensive research and testing, which also means extensive expenditure. And since Dutch broadcasters, Sound and Vision's primary client, do not use FFV1 either, there is little incentive to introduce it into the format mix.

All of this serves to underscore how deeply directional and historically oriented archival practices are. Changing an archive's path—and the formats that accompany it—is not trivial. New solutions to old problems are often difficult and expensive to reconcile with existing infrastructure, with the orientation and inertia an archive already has. Guidelines like IASA's *Handling and Storage of Audio and Video Carriers* speak in abstract terms about treating media files "like any other computer data."[38] But, as both Geoff Bowker and Lisa Gitelman have argued years ago, there is no "computer data" as such.[39] There is only data formatted in specific formats, governed by format standards or format conventions.

Nevertheless, FFV1 and Matroska have clearly arrived as preservation technologies that are now perceived as legitimate by many memory institutions that are actively implementing them in preservation practice. And this historic turnabout is being catalyzed by the work of the CELLAR group.

Out of the CELLAR and Into the Limelight: Standardizing Matroska and FFV1

The lack of backing by an established standards authority has, as we have seen, paradoxically helped FFV1 and Matroska gain a small foothold in the audiovisual archive. These media technologies had been circulating around the web and were gradually welcomed in some institutions without ever having undergone formal standardization. But over time, it became clear to some preservationists that, for these formats to find solid purchase in libraries and archives, they would need to be standardized by a body that the worlds of cultural memory *and* broadcast/Hollywood recognize as an authority in digital video technology. And since 2014, there has been an effort to create a completely open-source video digitization standard for cultural memory institutions using the Matroska container and the experimental FFV1 lossless video encoding. Yet again, the function of format standards as boundary objects mediating between communities becomes apparent.

WHAT IF STANDARDS WERE OPEN SOURCE?

The first major push for standardization evolved out of PREFORMA ("Preservation Formats for Culture Information/e-Archives"), a large-scale European project involving numerous educational and cultural heritage institutions that sought to allay some of the chaos of digital archival formats and shape future memory standards by providing authoritative reference implementations for common file formats. We could call this process *meta-standardization*: standardizing how standards are implemented.[40] Under the leadership of the Swedish National Archives, PREFORMA selected the FFV1 and Matroska video compression and container formats as its target case studies (along with other file formats for still images, documents, and audio).[41]

One of the applications that materialized out of this project was MediaConch, a tool that checks conformance to a particular standard. This allows the user to see if, for example, a TIFF image truly conforms to the TIFF specifications. Throughout the course of the MediaConch work, it became apparent that FFV1 and Matroska had such loosely defined specifications that a conformance checker would have a difficult time approving these kinds of files. The MediaConch team realized that for their work to be effective, standards would need to be developed for these technologies, whose completely open-source nature meant that existing files could not be expected to have much conformity.[42] This was the genesis of CELLAR.

The PREFORMA project concluded in 2017 but the work of standardizing Matroska and FFV1 carries on under the auspices of the IETF, the open standards organization responsible for Internet protocols. CELLAR (Codec Encoding for LossLess Archiving and Realtime Transmission) is the working group within the IETF tasked with developing specifications for the formats (Table 3.1).

According to the website for its email communication list, CELLAR's purpose is to "work to formalize specifications of the Matroska media container, FFV1 codec, and FLAC [audio] codec, based on existing work from the respective communities."[43] Ashley Blewer explains that the major

Table 3.1. How Matroska and FFV1 moved from their niches into libraries and archives

MATROSKA CONTAINER FORMAT		FFV1 VIDEO COMPRESSION CODEC
Developed by Steve Lhomme	**2002**	
		Created by FFmpeg as experimental lark
Becomes ubiquitous online as free format for video piracy in early 2000s		
	2007	Deemed most balanced and flexible in codec comparison by Vatolin et al.
Matroska-derived WebM format released		Austrian Mediathek starts use
Indiana Univ. starts use in digitization		Indiana Univ. starts use in digitization
	2012	
		BBC starts use; release of FFV1.3
PREFORMA launch		PREFORMA launch
CELLAR is formed, begins standardization		CELLAR is formed, begins standardization
	2017	
First IETF draft published		First IETF draft published
First mention in IASA guidelines		First mention in IASA guidelines
Flemish Institute for Archives announces migration away from MXF/J2K to MKV		LoC marks format as "acceptable"
Indiana Univ. completes mass-digitization, widely employing the container		Indiana Univ. completes mass-digitization
	2022	CELLAR submits v.4 specification to IETF
LoC adds MKV/FFV1 combination to its Recommended Formats Statement		LoC adds MKV/FFV1 combination to its Recommended Formats Statement

difficulty here is that both FFV1 and Matroska have existed as homemade files "in the wild" for over two decades. Such files, circulating around in a wealth of different "flavors," could not be counted upon to rigidly conform to any published standard:

> There are open source implementations of both formats [. . .]. However, there are concerns about the sustainability and credibility of existing specifications for the long-term use of these formats. These existing specifications require broader review and formalization in order to encourage widespread adoption.[44]

CELLAR aims to create documentation that can harmonize future implementations of these existing technologies to an agreed-upon template, developing standards for high-quality, losslessly compressed video digitization that parallel the quality of the Library of Congress's MXF/JPEG 2000 specification but without the reins of the commercial media industry and the prohibitive costs of implementation of AS-07. Much like the Library of Congress's AS-07 specification, CELLAR also holds the interoperability of files and software as one of the goals for its standards (Table 3.2).

At the file-production level, as sociologist Vincent Mosco notes, the distinctions between producers, distributors, and consumers can become indistinct.[45] Media scholars have studied this process and the resulting "convergence culture" comprehensively,[46] but there may still be scholarly blind spots in regard to how much media formats, specifically, both propel and irritate the blurring of boundaries between these communities of practice and their ability to actively participate in media culture. Open-source formats can, of course, develop a democratizing momentum: that is what leads to both film pirates and film archivists becoming interested in using the same format. But the flipside of egalitarian ease of access is the proliferation of loose, colorful format variations. Those are not especially problematic in the shadow economies of the moving image but represent a major potential point of failure in an archive.

Table 3.2. Overview of differences between the AS-07 and CELLAR guideline/standardization initiatives

AS-07 (MXF/JPEG 2000)	**CELLAR (Matroska/FFV1)**
Guideline for creating moving image files for long-term archiving	Working group within the Internet Engineering Task Force
Adapts SMPTE's MXF container and ISO's JPEG 2000 lossless video compression codec	Standardizes format specifications for Matroska and FFV1
Developed by the US FADGI with the aid of AMWA to support digitization efforts at the US Library of Congress	Continues some of the earlier work of PREFORMA, a European format consolidation initiative
AS-07 itself is freely accessible; the underlying MXF and JPEG 2000 standards documents are fee-based	CELLAR specifications and standards documents are available online at no cost
Developed with considerable input from American cinema and television broadcast producers	Developed with considerable input from European and American archives/libraries
Name changed to Registered Disclosure Document (RDD) 48 and published by SMPTE in 2018	As of 2024, in the process of being formally standardized at IETF

CELLAR is attempting to create coordinative standards from the loosely interpreted FFV1 and Matroska specification. This term, as used by Susanne Schmidt and Raymond Werle, refers to standards that are voluntary, self-enforcing, and coordinate (rather than coerce) relationships within a *system* of technological artifacts, rather than regulating stand-alone artifacts only.[47] In the case of the CELLAR standards, the interoperability of files and software within and between repositories is the object of this coordinative function. The CELLAR group believes that widespread adoption in the world of archives—and thus, by extension, format sustainability—can only be achieved by making formats credible via standardization.

Standardizing an audiovisual format via the IETF is an unconventional move, reactions to which have been mixed in the preservation community. The traditional global custodians that convey legitimacy to moving image standards are bodies like ISO, SMPTE, or IEC. These, however, often restrict access to documents via paywalls, effectively placing them out of reach for many archives.[48] This "runs counter to the ideals of the PREFORMA project."[49] One of the underpinning ideals of PREFORMA was the belief that digital preservation is best performed with well-described and adaptable technologies. Paywalls were seen as an obstruction to this because they make standards documentation (and thus the recipe for making files and encoding/decoding solutions) closed to anyone who cannot afford it.[50] This is not the case with the IETF, whose standards can be accessed free of charge and whose members do not pay membership fees and participate as individuals, not representatives of corporations.[51]

There was initially mild disagreement within CELLAR. Some of the group members questioned whether an Internet-standardizing body would understand the needs of archives.[52] Such questions of turf and territory carry extraordinary weight in standardization because they can permanently alter the course of technological development. The history of the Internet itself is the perfect example. Some of the earliest 1970s proposals for unified principles for the design of interconnected networks came from the International Network Working Group, an informal body of researchers from Europe and the United States. But their efforts quickly collided with the International Telecommunication Union, which made it clear that it alone was responsible for standardizing telecommunication networks.[53] Each of these groups favored a different approach to network design. We can only speculate how our world would look today if the X.25 network protocol had been standardized based on packets, as the Internet is today and as the Networking Group had advocated for, instead of the circuit-based standard ultimately adopted by the ITU. The former prioritizes flexibility, the latter, control.[54] Had things gone differently, we may perhaps have seen an Internet emerge twenty years sooner. CELLAR's position as a group of archivists and coders embarking on "their first IETF experience" thus, in part, recalls the early days of Internet standardization and the tussles surrounding jurisdiction.

IETF standards, however, are not only free. They also do not have any licensing barriers and standardization work within the organization is transparent and well-documented, which cannot always be said for SMPTE or the ISO.[55] The IETF standardization is unprecedented, because the process is centered around the needs of heritage institutions rather than hardware manufacturers or the film and broadcasting industries, and heritage institutions have historically rarely had a say in standardizing the formats that enter their custody.

The choice of the IETF did ruffle some feathers. One of our interviewees, Lars Gaustad, referred to the IETF as an "outskirts" organization, outside the usual media standards-generating world. James Snyder wondered aloud in our interview why the CELLAR group chose not to use JPEG 2000 instead of FFV1 to begin with, emphasizing that JPEG 2000 is an open specification, too, and anyone can write tools for it. Snyder worries that developing FFV1 and Matroska outside of the tried-and-true SMPTE domain could lead to interoperability problems with established production and post-production technologies:

> ... There is no way to take an FFV1 Matroska and put it into an AVID media composer. You can't do it. It doesn't mean it couldn't be done, but it means developing cost, working with vendors, all the stuff that we have already done for JPEG 2000, MXF, IMF, VCP's, all the other things that we've done. It means that the people who are doing content preservation are not working with the industry to make sure the industry can use their content if they ever want to [...]. It means literally the only way to view that content is on a computer platform.
>
> And if that computer platform cannot talk to other file formats, cannot be transcoded, cannot do the things that industry needs to work with their content, it's a dark archive. Period. I'm sure it's frustrating for folks who have made a decision on the FFV1 direction to have someone like me come in and say: 'you realize there's another industry here, in case you guys haven't taken into account?' And they, for whatever reason, didn't feel they needed to. But the fact is, at some point in time people are going to start asking for content and it's not going to be available.[56]

Furthermore, Snyder claims that CPU power has increased enough over the years to make the computational load of compressing and decompressing JPEG 2000 files negligible. Ultimately, he sees CELLAR's choice of

FFV1 over JPEG 2000 just as an example of the open-source community consciously refusing to align with industries like American television broadcasters and Hollywood.

Even though some audiovisual preservationists—particularly those who come from Hollywood or broadcasting—may look at CELLAR askance, the preservation expert and CELLAR member Reto Kromer emphasized that he is proud of the fact that this standardization effort, unlike any audiovisual preservation standards in the past, is largely taking place "by and for archivists." Many of our interviewees cite the participation of archivists as a chief reason for their interest in observing or participating in the CELLAR work. Furthermore, CELLAR's choice of the IETF may be a signal that, for many institutions, the tide is turning from industry standards toward open-source technologies for video digitization.

Despite the controversy, the IETF has, in fact, seen some library and archives collaboration over the years. For example, the BagIt File Packaging Format, which is used very widely in digital preservation work in libraries and archives around the world, is an IETF 1.0 specification that was developed with significant input from the Library of Congress.[57] Other time-based media standards like the OGG audio container format, VP8 video format, and the Opus audio format, are also open and in various early stages of candidacy for becoming full-fledged standards at the IETF. Lately, there has been a boom in the development of such open-source, do-it-yourself preservation and digitization solutions, such as software for digitizing video, repositories for packaging digital content for storage, metadata harvesting tools, and so on. In other words, the tools, methods, standards, and specifications governing video digitization are becoming more participatory, with much more buy-in by the people doing the cultural memory work.

The social character of this undertaking is underscored through collaborative events like the No Time To Wait conference, an avowedly open source-aligned event that encourages collaboration and knowledge sharing between participants. New formats and technologies are often presented at this annual venue, and the CELLAR group has had a prominent presence there since the conference's inception. The conference title emphasizes the audiovisual preservation community's need to stop waiting for large actors to develop standards and best practices and instead take matters into its own hands.

PURELY PRAGMATIC OR A MATTER OF PRINCIPLE: WHY SOME ARCHIVISTS OPT FOR OPEN SOURCE

For some institutions the choice to use open-source technologies—and for our purposes, this simply means that the code for a technology is openly available without a paywall—is pragmatic. FFmpeg, FFV1, Matroska, and other open-source tools are used in audiovisual preservation shops large and small, for-profit and nonprofit. Indiana University's decision to use FFV1 on a massive scale was largely driven by the "it just works" ethic of both the university and its digitization partner Memnon. In fact, as Andrew DaPuzzo noted, Memnon's parent company Sony never questioned its decision to implement FFV1 nor Matroska, despite its association with piracy and Sony's earlier resistance to supporting the container in its early Blu-Ray players. As DaPuzzo put it:

> Never had the MKV been an issue for that because I've dealt with torrents before. I was at Sony and, in fact, working in New York City, and even then BitTorrent [wasn't] always used for piracy. So I had the misfortune to be working with a book publisher, a large one that had used a torrent-like file delivery system and within, I would say, 10 minutes of opening that file, I had half a dozen Sony security people visit me in my office. Because they had detected through their firewall that this file comes through and rather than block it, the position at the time was, we're going to go to the person who downloaded it and find out what's going on and have a very serious conversation. So Sony takes this, to be clear, very very seriously. But never an issue with MKV.[58]

According to John Dunn, Assistant Dean for Library Technologies at Indiana University, the format choice at Indiana was less about the ultimately political question of open versus proprietary than it was about documentation access and the potential for viable playback and encoding tools, as well as quick and easy support from the growing user base.[59] When asked why CELLAR chose FFV1, FLAC, and Matroska to standardize, Ashley Blewer also found the openness of these technologies to be key to their usefulness in the preservation:

> I will be a preservation evangelist here! The obvious, simple answer is that they are open formats and so you're never locked out of having the knowledge to work on these themselves and they were picked for that reason. When we decided to go with IETF to standardize, that was a big important part of that too. So that would be open forever. For me that's the fundamental—there are adjacent reasons why they're good as formats, but overall it's because they're open that makes them such a good fit for preservation.[60]

While pragmatism, usually born out of staff, infrastructure, and funding considerations, tends to sway the format standard choice in libraries and archives, there appears to be a growing number of professionals who choose to align themselves with open source for ethical reasons. With respect to Indiana and Northwestern University, both institutions, according to Josh Yocum and Mike Casey, have a "culture of openness" that helped FFV1 and Matroska easily find purchase with IT staff and administrators. In other cases—such as the City of Vancouver Archives—that culture can be bolstered through legal requirements and guidelines.

Many of our interviewees expressed support for open-source technologies on ethical grounds. This support was often articulated as a conscious disavowal of the SMPTE-aligned television and film production industries. While their own decision in this matter was largely determined by technical factors, in our interview at Scenesavers, Jim Schupert referred to the divide between adopters of MXF/JPEG 2000 or Matroska/FFV1 as a "North vs. South" divide, implying that it can be very contentious. This sense of contentiousness also underpins Ben Turkus's comment about taking a stand for open source. Openness and sustainability are so commonly intertwined in the literature of digital preservation that, to some practitioners, they are the same. As Lars Gaustad put it, he felt obliged to make the move to the CELLAR standards "because of the commandments of digital preservation."[61] Hermann Lewetz voiced a frustration that was registered often at the No Time To Wait conference when he said that he is tired of broadcast archives behaving like they are the "normal" archives by speaking for nonprofit institutions like the Mediathek—institutions that have workflows and obligations very different from those of broadcasters.

It is in discourses like these that formats can help us dislodge a conceptually stale notion of the archive. As the film historian Erika Balsom recently pointed out in a brief appeal, scholars of film and media have a habit of interpreting the word "archive" as a "singular definitive," letting it serve as a generic stand-in that flattens out geographically, politically and pragmatically entirely incongruous institutions.[62] (Much like the fuzzy collective term "memory institution," whose use the museum scholar Helena Robinson has criticized on similar grounds.[63]) Often enough, that generic image of "an archive" is modeled after the large, comparatively well-financed and technologically well-equipped archives of Europe and the United States. In their contentiousness, format standards exert a pressure to consider the morphological and epistemological differences between what may superficially be subsumed under the broad label "archival practice." Acknowledging these finer gradations might aid us in enunciating a more balanced scholarly understanding of global archival politics. Recent debates at preservation conferences show that archivists and scholars are increasingly concerned with the implementability of preservation standards across the world as well as their environmental burden. As the film archivist Carolina Cappa asks: "Do we use the technologies created in the Global North, whose operation is scientifically proven although out of reach of our limited budgets and sometimes not very useful for our needs, or do we create our own experimental tools?"[64] But let us delve into this point in more detail in the next chapter.

Not all broadcasters take issue with open-source technologies. WGBH, one of the premier public broadcasting stations in the United States, has recently adopted FFV1 for its preservation digitization work. While the attitudes of some broadcasters may rankle archivists like Lewetz, there are many who are in alignment with the Austrian Mediathek in terms of willingness to implement open-source standards. Lewetz noted with some vindication in our interview that, "I proved FFV1 is a serious tool for archiving."[65]

Jean-Christophe Kummer advises many archives and libraries around Europe on digitization. He observed that he finds the most vocal open-source evangelists to be younger professionals who started their careers after 2000 and come from a "geeky point of view" of self-taught, motivated coders who prefer to find their own solutions to problems rather than wait for large developers like Apple and Microsoft or standards bodies like SMPTE or the MPEG.[66] While many of our interview subjects have clearly well-informed technological, infrastructural, and pragmatic reasons for their preference of FFV1 and Matroska, they continually made reference to their allegiance to the ethic of freely available code.

The Appeal of Open Source: Agency, Transparency, Customization, and Affordability

The black-boxing of the MXF and JPEG 2000 standards, the potential for vendor lock-in, the lack of AS-07 compliant tools, and the byzantine complexity of the standards have combined to pave the way for the CELLAR standardization work. This informed Ben Turkus's comments about Matroska closing the gap between large and small institutions. This attitude is shared by several of our interviewees who are CELLAR contributors as well as preservation specialists at the New York Public Library, Indiana University, and others.

The CELLAR standardization work is an attempt to give archivists and librarians more agency in the development of digital video standards than they have ever had before. The CELLAR standards documentation is also freely available to anyone with an Internet connection. This strict open-source policy, followed since the days of the PREFORMA project, is a conscious signal that archiving, whether institutional or private, does not need to rely on black-boxed machines and software inherited from the media industries, whose functionality is pre-determined by vendors' economic interests and planned obsolescence.

Freedom from proprietary and vendor-dependent solutions serves as a preservation strategy. Since the "spare parts" used in the creation of the format are open, playback software can also, in theory, be adapted for future hardware, other processor architectures, or as yet unknown contexts without the need for reverse engineering or other "hacks."[67] An open and transparent approach to format standardization frames archiving as a collaborative activity, and open tools make possible new subjectivities of archivists as tinkerers. Critical policy framework decisions based on the democratic participation of many institutions will also ultimately benefit the end users of the archives, including media scholars and historians.

The CELLAR group's work has helped to pull FFV1 and Matroska out of *sub rosa* niches and into more mainstream appraisal and use. The inherent affordances of these technologies made them amenable to the library and archives community—a community that is now, through the auspices of the IETF, conforming them to standards that will aid their use and adoption in memory institutions.

There are other important reasons for some of our interviewees' interest in CELLAR: Ben Turkus cited the fast, clear responses he receives to his questions on the CELLAR email list.[68] Brecht DeClercq, then preservation specialist at the Flemish archiving institute VIAA (today part of Meemoo, the Flemish Institute for Archives), cited its nonprofit status and the openness of the standards they are producing as a potential corrective to the periodic difficulty finding support for tools that fail due to market pressures, what he calls the "ugly side effects of capitalism" in preservation.[69] Dave Rice noted that the lack of barriers to access to the CELLAR work means that more archivists will feel empowered to participate and influence the design work in favor of the needs of the preservation community instead of for-profit industries.[70] It is also an express goal of CELLAR to create documentation that is both technically precise but also readable by people in the preservation community, which is not always the case with SMPTE or ISO standards. This level of clarity is not always achieved by the more intensively engineering-specific conversations that play out on the email list, which is a chief reason Jim Schupert of Scenesavers claimed to be a "lurker" in that space—watching, monitoring but not often participating.[71]

Despite the many recognized advantages, the degree of involvement in open standards development can itself be a contentious catalyst of friction. Peter Bubestinger-Steindl, for instance, expressed some hard feelings about Indiana University's use of open-source tools without investing heavily in their development:

> [Indiana University's adoption] was a major breakthrough for the perception of FFV1, just because they put money where their mouth is. [. . .] This is very helpful for our community—to have an argument that says, okay now can we please get over this 'nobody uses it and it's only a cheapo thing' because they're serious about it? The bitter stuff is: I have not seen a single contribution from them, neither in time nor financially, to what we did, in order to make this format happen and keep it alive and so on and so forth. And that's actually what I found a bit sad. But maybe they did and I just didn't see it.[72]

This frustration exemplifies the ongoing problem of the "free rider" in digital preservation—institutions using resources without replenishing them. While a standard is, technically, a non-rivalrous technology (that is, it can be used over and over without being used up) Bubestinger-Steindl's concern is that the labor and resources needed to develop and adapt a standard are not infinite. He is troubled by Indiana's high-profile use of FFV1 and Matroska without being comparably present and active in the CELLAR development work.

Ashley Blewer asserted that the fact of FFV1 and Matroska being standardized by a reputable body like the IETF can increase the support of administrators who might otherwise be unfamiliar with them. As she says, standardization "helps in terms of anybody trying to prove things to their bosses that don't understand this but need the buy-in from someone else."[73] Furthermore, the increasing use of FFV1 by broadcasters in Europe is, according to Brecht DeClercq, reason to hope for a cascade effect in the encoding's adoption across the continent. And indeed, Marion Jaks of the Austrian Mediathek observed far less resistance to FFV1 as of our interview in 2017 than her colleagues did initially in 2010.

The video formats Matroska and FFV1 are both symptomatic and exemplary of a larger turn toward open-source technologies in the archive. Memory institutions are increasingly shrugging off the path dependence of industry-preferred standards and work with software like FFmpeg or VirtualDub, another transcoding tool that includes support for FFV1. The openness of VirtualDub (licensed under the GNU General Public License) is what motivated Erik Dix, archivist at the Notre Dame University Archives, to use it. And its integration of FFV1 allows his institution to digitize video at high quality while saving storage space.[74]

As Langdon Winner points out, sometimes technologies are adopted simply because of their compatibility with other technologies.[75] Early adopters like the Notre Dame University Archives are attracted to video formats that can be produced with freely downloadable, near-ubiquitous software tools. Erik Dix said that he uses the FFV1 encoding to losslessly compress video for storage because the encoding works seamlessly with VirtualDub, the software they use to digitize their analog originals. The compatibility is not perfect, however: VirtualDub does not easily create or decode the Matroska container, so Dix wraps his FFV1 video essences in the Microsoft-developed AVI file format.[76] The Notre Dame Archives, like many similar institutions, operates with considerable budgetary, staffing, and resource constraints that force them to use inexpensive, user-friendly tools like this to do their digitization work. Thus, as Winner suggests, FFV1 and Matroska more easily find purchase in these institutions if they are compatible with existing technologies that are cheap and easy to implement.

The broadcast and cultural heritage domains have quite different views on audiovisual preservation. The former privileges speed and efficiency, while the latter is often willing to sacrifice both in favor of high image and sound quality and faithfulness to original recordings.[77] Thus, tools like VirtualDub, which Dix admits can be cranky at times and in need of patience in updating, patching, and retrofitting, can find a better home in libraries and archives than in fast-turnaround, high-stress environments like TV production and distribution. Because their code is open and freely accessible, tools like Archivematica, VRrecord, FFmpeg, and VirtualDub allow for a kind of transparency and ease of "getting under the hood" to adjust for local practice. This responsiveness to an institution's idiosyncrasies, coupled with direct, quick access to the tools' coders, can make these techniques more appealing than the costly, "black-boxed" technologies used by media producers, editors, and post-production houses. FFmpeg is only available under a copyright-free license—you would not be able to pay for it even if you wanted to—and its website has an active forum that encourages information-sharing among users.

The cost of tools is, of course, also a factor in their adoption. As Peter Bubestinger-Steindl and Marion Jaks said, JPEG 2000 workstations can be very costly to build. Compare these costs with the cost-free FFmpeg, VRrecord, or Virtual Dub software and the appeal becomes apparent. The sharp disparity in cost has been felt by software manufacturers. Furthermore, the ease of access to open-source tools has encouraged libraries and archives to actively participate in their development and adaptation. As Lars Gaustad noted, the slow progress of AS-07 has been a factor in its equally slow adoption, and of Matroska and FFV1 overtaking it. As Gaustad put it:

> Even though Matroska and FFV1 aren't properly standardized yet, they are moving at the speed of light compared to normal standardizing work I think. I've been involved with the standards in the Audio Engineering Society back in the '90s and early 2000s and it took us forever because one of the problems is it's voluntary work. I mean you can always have your employer support you in it, but still it's voluntary work and you may meet twice a year so it takes forever to get results. That's been a bit frustrating and I think the time scale of standardizing FFV1 and Matroska is impressive.[78]

Steve Lhomme told us that his work developing Matroska was done in his spare time and progressed slowly. In comparison, the CELLAR group has been able to marshal more resources than he could, and the format's conformance to their nascent standard can move briskly as it nears completion.

OPEN SOURCE REMAINS AN UPHILL BATTLE

While the new, open-source mode of developing and standardizing media digitization technology has been advancing, it has not been without resistance from entrenched communities. The 2010 presentation by staff from the Austrian Mediathek at the joint AMIA and IASA conference—in so many ways a central dramatic event in the narrative of our book and a watershed moment for audiovisual preservation at large—revealed a resistance to the FFV1 codec in particular and non-standard video formats in general, partly because of historical understandings of what it means to be an archivist. Those ideas are not easy to change. Peter Bubestinger-Steindl, who attended the conference session, recalled that people in the SMPTE-standard camp called the Mediathek presenters "open-source weirdos" and "reckless" and "unprofessional" for their choice of FFV1.

> [T]he head of the Mediathek came home from a conference a few months later and said: "the people are shaking their heads [at] what you guys are doing, I don't know what to tell them. I don't know if it's right what you are doing, what the hell?"[79]

Ben Turkus attributes these differences in opinion to a generational divide he observed at AMIA conferences, one in which youth aligns with open-source and age aligns with SMPTE/MPEG/ISO.[80] Turkus claims that this divide can partially be explained by what he calls a "herd mentality" within the audiovisual preservation profession, wherein professionals from institutions of similar profile tend to align their preferences with respect to standards, hardware, and software. And indeed, we have seen evidence of this in other conversations, too. Marion Jaks of the Austrian Mediathek explained that her institution initially chose AVI as a container format because it was a politically safer choice. They already knew FFV1 was controversial as a compression format in 2011 and were reluctant to dabble with Matroska as a container, partially because they were already getting pushback from the community and because MKV was a well-known piracy format.

While the adherents to a particular standard usually have clearly defined reasons for their allegiance, among our interviewees there was a clear preference for well-adopted standards, which could be explained, in part, by a desire to align with libraries or archives of similar profile. This is a key point in the political economy of format standards: Mosco argues that social relations organized around power, even in the face of resistance, can be a strong motivator for choice of technology.[81] Turkus's comments point to an audiovisual preservation world wherein adoption of technology is not just motivated by a cold, rational analysis of affordances and constraints, budgets and "full-time equivalents," but also by social ties and the prestige of particular professionals and institutions. Peter Bubestinger-Steindl finds that, in attending conferences like AMIA and IASA, some people do not feel comfortable discussing these issues because it is easy to be stigmatized by other peers for aligning with a particular format:

> A lot of people are reluctant to engage in a public discussion about this because, especially when they come from the preservation side, they feel like they're not knowledgeable enough on the details of digital image and media that they are a bit afraid.
>
> They're like, "oh I don't know, maybe people will think I'm stupid or I'll say something and it makes me look bad." That's one thing. And the other thing is that some people are scared coming up because they might ask the wrong questions, but because it could be seen that they have this opinion on the subject and as I told you before, it's not only a technical thing but also a political thing.[82]

Nonetheless, after he presents at conferences, Bubestinger-Steindl admits that he is usually bombarded by one-on-one emails from people who are interested in knowing more about these tools.

The lack of standardization, as Ashley Blewer noted, is a common argument against trusting a container or encoding and part of what CELLAR is trying to overcome. Adherents to the value of standards can passionately defend this logic, which can lead to the quasi-religious timbre to some discussions related to standards choice, according to Brecht DeClercq. In his interview with Jimi, Mike Casey mentioned that this lack of fully formed standardization caused one prominent video engineer and digitization sage, Jim Lindner, to publicly question Indiana University's choice of FFV1 and Matroska.[83]

Even in the world of open source, standardization still conveys some safety. More rigorous testing of the encodings and containers will likely be needed to sway more potential users. James Snyder, for instance, called the lossless compression capabilities of FFV1 into question when he asserted that because the format uses interframe encoding—wherein picture data is not encoded in discrete frames but instead is predicted between key frames—it cannot truly be called lossless. The interframe nature of the FFV1 codec may haunt its reputation. Regardless of the "speculative" nature of FFV1, the resistance to it and Matroska have been breaking down for many archives and libraries, particularly those with a pre-existing culture of or affinity toward open source. In his audiovisual consulting business, Jean-Christophe Kummer believes that it is the older, established archives that may resist newer formats, particularly if they have already sunk considerable costs into using something else. He has even found that any stigma regarding Matroska and its association with piracy has, for the most part, dissolved with his clients.[84]

The do-it-yourself patina of open-source technologies does not wash off easily, however, and there is a lingering belief among archivists that technologies that do not arrive fully formed from well-known industry may require tinkering to implement. In the words of Andrew DaPuzzo of Memnon, "Open-source isn't free like 'free beer,' it is free like 'free puppies'—it can cost you later."[85] Institutions without the resources to allocate for trial and error—and, as we have seen in all of the previous chapters, media preservation involves *a lot* of trial and error, all the time—may have little inclination for technologies that may require retrofitting down the road and thus may find industry standards, with their luster of inevitability and completeness, more appealing.

CONTESTED VISIONS OF THE FUTURE OF "CONTENT DELIVERY"

There is an irony here in that the MXF container was designed to deal with a wide array of potential content distribution channels,[86] yet Matroska and FFV1 now appear to be better equipped to handle this proliferation than their older, industry-designed counterparts. FFV1 is poised to find broader purchase outside its current niche. Reto Kromer is working to expand the FFV1 standardization beyond video into photochemical film digitization. Marion Jaks said she has seen a rise in acceptance for the Mediathek's choice of the encoding.[87] And, perhaps most compelling is an apparent acknowledgment of FFV1's utility as a high-quality, long-term preservation encoding by Jim Lindner, who had been a vocal critic of the Mediathek's presentation in 2010. In a recent post to the AMIA email list, Lindner had this to say: "The great thing about soft formats like JPEG 2000 and FFV1 is that they can be readily documented in human-readable form on a long-lived medium. Any competent computer programmer could figure out how to read it and transcode to whatever the current preferred format is."[88]

The stabilization of a technology is not a singular event. It is, instead, a process that often takes years.[89] CELLAR is working to stabilize FFV1 and Matroska by standardizing them—creating clearly defined and widely agreed-upon documentation that will confer upon these "wild" technologies a status that the preservation profession can deem trustworthy. Standardization will almost certainly help FFV1 and Matroska to proliferate. As of 2023, the standardization of FFV1 and Matroska has reached a plateau. While their adoption by American universities and European archives and libraries continues, it is unclear if Hollywood and American television broadcasters will ever implement them on a large scale. Ashley Blewer expects that

> film/broadcast won't choose the CELLAR standards because they are working with whatever the latest raw formats are, and I think we'll see more of original formats without reformatting rather than additional conversion steps. I see FFV1+MKV as an archival format, and especially a format for transferring media off of analog tapes and being welcomed into the digital realm for the first time.[90]

Broadcasters require technical infrastructure and knowledge not just to preserve files but to play them out over the air as well, both over terrestrial (over-the-air) and cable transmission. Because media companies like NBC are geared for MXF and MPEG playout—and tied to other, international format standards for digital television broadcasting—they are, at least for the near term, unequipped to broadcast files in other formats.

Open-source standards are relatively easy to play back on computers that have tools like VLC Player. However, as James Snyder rightfully objects, it is one thing to be able to play back content on a computer, but another entirely if a broadcaster needs that footage someday. As quoted earlier, through the course of our conversation, James theorized that many implementers and supporters of FFV1 and Matroska, because of their library and archives backgrounds, are thinking of the content in their care as likely to be delivered over computers in the future, rather than over broadcast networks. This conversation illustrates that the differences between open-source and industry-developed standards adopters are not a nonprofit versus for-profit distinction. It is rather, at least partially, a difference in vision of the future of media and of how the materials in an institution's care are likely to be disseminated. Open-source adopters tend to anticipate their media being distributed over computer screens, while industry adherents continue to see cinema and television screens as the primary distribution vectors for their content.

The time and resource constraints that librarians and archivists tend to work with are, furthermore, pushing them to focus on the near-term goal of harvesting content from decaying physical carriers. This, too, can be an incentive toward the pragmatic, open-source mindset. While experts like James Snyder understand the resource scarcity that smaller institutions are working with, they are still worried that this content will be difficult to act on later. It is likely the current path dependence of broadcasters and cinema producers, with their investment in digital cinema specifications that rely on MXF and/or JPEG 2000, will continue to resist FFV1 and/or Matroska, even after the CELLAR group finishes its work.

Conclusion

Trevor Pinch, in writing about the adoption of the Moog synthesizer in the mid to late twentieth century, argues that the instrument's salespeople played a critical role in creating a market for it. With its wires and electronic patch-bay appearance, the early iterations of the Moog did not look very much like an instrument. It was a hard sell until the release of the MiniMoog, which had a keyboard and a much more comforting, musical appearance.[91] Lewetz's presentation at the 2010 conference served a very similar function: that of a salesman trying to convince a skeptical audience of the utility of a technology that did not have the appearance of a preservation video format. Unbeknownst to its early developers *and* skeptics, FFV1 was inscribed with affordances that primed it for use in libraries and archives.

Matroska's adoption in cultural memory institutions was similarly slow. While it proliferated widely in online piracy, it lay fallow for several years out in the "legitimate" media worlds until 2010, when it was adopted as the WebM container, which is used in HTML5 video and audio elements that allow browsers like Firefox and Chrome to natively decode and play back time-based media. This mainstream implementation was one in a series of events that blur the line between for-profit and nonprofit actors in the history of media formats, and a story to be told in the next chapter.

One of the chief reasons for the choice of Matroska for this purpose was its royalty-free nature.[92]

As we have documented, both FFV1 and Matroska have begun emerging out of their shadowy niches into the sunlit world of international cultural memory digitization and preservation. Part of the reason for this is the nature of technological development which, in the words of Lucy Suchman et. al, involves ongoing appropriation and refashioning of existing technologies.[93] Like the Moog salespeople in Pinch's study, Bubestinger-Steindl, Lewetz, and others had to make a market for FFV1 by refashioning and articulating it as a viable technology even though it looked and felt different from the standardized tools that came before it. The Austrian Mediathek's presentation helped to expose the "market" of audiovisual repositories to the notion of using FFV1.

And while Matroska was not designed specifically for cultural heritage or media preservation, its design was elastic enough to be amenable to this kind of use. Since it can bundle almost any kind of content and bitstream together and because it is freely available without cost, the format has comfortably situated itself in both illicit practices of piracy, as well as the budget-constrained world of libraries and archives. Its support in Blu-Ray players and in a range of free and popular digital media players has helped it move from piracy into "legitimate" content consumption.[94] WebM and piracy gave Matroska a foothold in the dissemination of video content online.

Clearly, FFV1 and Matroska are technologies that can move between sociotechnical worlds. Neither of these developments, however, caused very many American memory institutions to take a chance on these formats. It is far safer, politically and technologically, for archivists and librarians to adopt tools and techniques championed by large institutions with high name recognition like the Library of Congress than argue for esoteric tools like FFV1 or Matroska, the merits of which may require extensive understanding of video engineering to explicate. Video digitization and preservation in libraries and archives can be a bit like a pool party wherein participants are willing to dip a toe into the water, but no one wants to jump in until they see someone else safely do it first.

Video digitization and preservation is a resource-intensive, time-consuming, and costly endeavor, and choosing compression/container combinations is not done lightly by most institutions. This is because a file format or encoding that, for whatever reason, might not be easily decoded in the future, can lead to costly re-digitizations or reformatting at best and, at worst, the complete inability to access that material because of age-related failure or obsolescence of the original. And without the compulsatory power of a standards body or patent reinforcing the format, who could say what was or was not a properly formed Matroska file? But while the free-range nature of early Matroska files has been vexing for the CELLAR group, it increased the user base and the interpretive flexibility of the format. This flexibility, coupled with the low barriers for entry for CELLAR participants, has helped Matroska and FFV1 to become boundary objects that have found homes in media production, piracy, and cultural memory preservation.

In studying the audiovisual formats and standards used in archives and libraries, we have found that MXF, JPEG 2000, Matroska, and FFV1 are not simply media technologies. They are surfaces along which group identities are formed, allegiances are made or broken, and ethical stances are negotiated. There are users who have chosen MXF and JPEG 2000 to align with Hollywood and television broadcasters. There are also users who have decided *not* to use these format standards *because* of their links to these industries. There are many users of the CELLAR formats who have decided to consciously align their institutions with the open-source community instead of the proprietary standards, for pragmatic as well as idealistic reasons.

While many CELLAR specifications adopters and contributors are simply looking for low-cost solutions that work for them, they are also contributing to a zeitgeist in audiovisual preservation characterized by increased enthusiasm about free open-source tools developed by and for the preservation community. This change in thinking goes along with a sense of increased agency among many of these professionals, as they no longer feel beholden to closed and proprietary "hand me down" standards and technologies from film and television. Some members of CELLAR argue that the openness and transparency of the standardization—the fact that the development work is accessible to all online—means that the nascent standards for FFV1 and Matroska will have greater utility to memory institutions because, in theory, they can see how the sausage is made and can contribute to that making freely and in real time.[95] This collaborative effort is perhaps *in parvo* a fulfillment of some of the old utopian ideals with which "cyberspace" was heralded in its early days: The online, open communication avenues can allow more players to speak and have input on the work—a transparency not possible in traditional, closed-door standards-making meetings that may require travel and membership costs. On a theoretical and historical level and from the perspective of format studies and the study of standardization processes, FFV1 and MKV are also striking examples of how amateur and seemingly illicit formats *can* enter formalized standardization processes through word-of-mouth but also through unconventional vectors like piracy. While MKV might be an unusually obvious example, it demonstrates yet again just how much of a political object formats and format standards are.

By contrast, MXF and JPEG 2000 have had trouble finding the interpretive flexibility needed to live comfortably in both the library/archives world and that of broadcast/cinema production. This owes at least partly to the closed nature of the SMPTE and ISO standards bodies, in which few librarians and archivists could participate. That lack of participation, coupled with the cost of the standards documents and encoders/decoders for JPEG 2000, can be significant barriers to entry for many memory institutions. Contrarily, the IETF has had a history of collaboration with librarians and archivists that SMPTE does not share. SMPTE, because its makeup consists largely of Hollywood and broadcast specialists and technicians, strongly directed the development of MXF toward the needs of those industries, as evidenced by the technical issues related to, for instance, interlaced video that became known during the early AS-07 specification work.

With respect to FFV1 and Matroska, it is too soon to tell, but many of the CELLAR participants believe that the IETF's transparency in the standards-making process will help to make the end users of these standards more aware of the work that went into developing them. This could lead to a considerable shift of power into the hands of smaller institutions within the archival and preservation community with respect to these technologies.

FFV1 and Matroska are coming of age in a time of cheaper and easier-to-access tools for creating and handling digital time-based media objects. The decreased cost of tools like VLC Player and FFmpeg as well as the free and open nature of Matroska and FFV1 have propelled their adoption by many memory institutions. It is possible that this adoption will cause a cascade effect—now that many institutions here and abroad have chosen FFV1 and Matroska, more will take the plunge. This means, as James Snyder says, that everyone working in large preservation facilities will have to deal with these products, like it or not.

Snyder noted that, because broadcasters in the United States have made up their minds about MXF and JPEG 2000, it is unlikely they will change any time soon. However, as more archives choose Matroska, FFV1 or their combination, content producers like NBC will have to be able to transcode and play out these formats. That means that these open-source technologies will, over time, insinuate themselves into sectors of industry that would not otherwise use them for long-term preservation purposes. But—and this is both an important caveat and one of the findings of our research—software manufacturers' support for these emerging format standards will be essential in weaving them into everyday practice. From the outside, the preservation of audiovisual heritage often appears as a one-way pipeline between two endpoints: producers, broadcasters, or collectors on one end, and archives on the other. But the objects that travel along this trajectory are greatly affected by the politics of software manufacturing and the materiality of the various digital format standards circulating across the tech industry, too. Many audiovisual archivists are likely to work with FFV1 and Matroska even if they cannot be easily generated with tools used by filmmakers and producers, but in order for those formats to stabilize as technologies and shed off their lingering affiliations with piracy, dominant video software manufacturers like Apple, Adobe, or Avid will need to be brought into the fold, too. The CELLAR group's lobbying efforts to get the encoder integrated into professional applications are documented in, for example, Adobe's online forums since 2015, and while tools like Premiere do not offer the possibility to export FFV1-compressed video natively as of 2024, workarounds are possible with the use of plugins. It remains to be seen if popular editing software will embrace these formats. Archivists would certainly welcome it. As Brian Wheeler of Indiana University put it:

> It's my hope that as time goes on and more people pick up both FFV1 and Matroska as a pair, that we'll start to see more support in editing tools. That the stigma of "oh, MKV is piracy format" will go away and Premiere will start to support it [. . .]. I think it's going to be a long long time, but I think it will eventually get there.[96]

One of the most important findings of this book has been the gradual change in attitude toward open-source technologies in libraries and archives. Where there was open hostility from some quarters in 2010 to the thought of using loosely defined tools, the work of CELLAR to standardize these tools and the work of institutions large and small, both in Europe and the United States, have changed minds about the utility and soundness of using them. This has also helped to break down some of the entrenched resistance to open-source tools and to change the mindset that it is somehow unprofessional to use open code instead of traditionally designed and disseminated specifications and standards. Jim Bird of Scenesavers comments that "[i]n terms of longevity it's like the genie's out of the bottle. You couldn't stomp it out, it's everywhere, everybody already has the source code and you know there's lots of people looking at a lot of details about it. [. . .] You either go with open and constantly evolving, that has levels, or lock it down with a SMPTE big long number."[97] The use and fostering of the development of open-source tools is, in fact, a recommendation to the digital preservation community by the US National Digital Stewardship Alliance.[98] Bird went on to say that while SMPTE as the audiovisual standards body made sense in the analog video days because of the need for playback decks and tapes to be interoperable regardless of manufacturer, they are no longer the only—or the most efficient—game in town, as evidenced by the speed with which CELLAR is working. As the lines between traditional brick-and-mortar archives and the Internet continue to blur, standards like FFV1 and Matroska, which are well-suited for modestly funded institutions that focus on free dissemination of content online, will likely grow in appeal.

A growing body of recent literature on archives has been dealing with amateur, informal, "accidental," "commons-based," non-institutional, and anti-institutional forms of archiving and preservation.[99] The promise of such alternative archiving practices is that they might support overheard, ignored, and suppressed forms of memory and enable more open, diverse, and just modes of remembrance than traditional archives have. In our view and based on the conversations we have had, all archiving is a highly enfolded field of practice in which established institutional archives and activist modes of preservation cannot be understood in oppositional terms only. Yet while it would be naive to assert that video formats like FFV1 and Matroska can directly advance the progressive politics of anti-institutional archiving, there is certainly an argument to be made that they have brought many "traditional" archives into closer proximity with a novel form of politics, enacted through open-source technology and open standardization. But, as we are about to find out in the next and final chapter, those format standards can also very effectively muddy the waters when it comes to distinctions like the one between for-profit business and nonprofit public work, and the "open source" label does not necessarily mean that the standardization process is really welcoming to everyone.

4

Standards at Work

For-Profit, Nonprofit, and the Global and Social Technopolitics of Standardization

When we embarked upon our research for this book, we articulated the work as an exploration of the sharp distinctions between the for-profit and nonprofit communities. Initially, our expectation was that the open-source video formats Matroska and FFV1 would have little, if any, input from for-profit actors. As it turns out, the story was far more complicated. It became apparent to us that this boundary is a shifting, porous, and hazy one at best, and insufficient as an explanatory framework. There is no clear dividing line between for-profit and nonprofit participation in the development of closed or open-source standards.

Film producers and distributors, television broadcasters, and nonprofit cultural memory centers all engage in media preservation, digitization, and dissemination. It is therefore hardly surprising that they would want to exert an influence on how these practices are performed. In this concluding chapter of the book, we first look at the remarkable interweaving of profit-oriented and nonprofit influences in format standard development. New collaborations between nonprofit foundations, tech manufacturers, and streaming giants are rewriting the rules of the game and signal a shift in the dynamics of standardization—not only of video formats but of technological standards at large.

In the quest to safeguard moving images, archivists face formidable barriers. The pursuit of "forever media" solutions, promising everlasting data storage, has proven illusory. The middle section of this chapter looks at some of the overarching constraints with which archivists contend. Time is one of them: material deterioration and administrative pressures leave archivists with little choice but to digitize, and lossless formats are not always a viable choice. Access to expertise and documentation presents another layer of complexity. The cost of accessing standards documentation can be prohibitive for many institutions, posing a barrier to adoption. Finally, our interviews show that even in a digital world, the physical proximity between archives and allied institutions remains critical to the functioning and adoption of formats. Navigating these temporal, financial, spatial, and expertise-related hurdles requires creative solutions and a flexible approach. Subtending these facets is the social dimension of the archival assemblage: trust.

In the concluding section, we are confronted with a web of issues that extend far beyond the technological—the gendered nature of video standardization labor and its role in neo-colonial power structures. This section situates our work vis-à-vis the stark underrepresentation of women in technical standardization. This gender disparity is not isolated; it resonates with the exclusion of voices from the Global South, as European and North American archivists exert disproportionate influence over global standards development. The chapter raises the provocative question: Can a video standard function as a colonizing technology? Increasing calls for equity and sustainability in preservation challenge us to consider the global implications of standardized video formats.

Observing these global and social dimensions of archival standards, this chapter provides a thought-provoking conclusion, reminding us that the politics of standards are deeply intertwined with gender dynamics, colonial legacies, and the power structures that shape our world. It underscores the need for a more inclusive, equitable approach to standardization and sets the stage for continued exploration of these issues in the preservation of audiovisual heritage worldwide.

Does "Open Source" Imply "Nonprofit"?

The genesis of the MXF container format is a good starting point to unravel the interplay between profit-driven and nonprofit actors in video standardization. As audiovisual preservation specialist Richard Wright noted in our email conversation:

> MXF was developed (so far as I know) for broadcasting, and Bruce [Devlin] was working for a company (Snell & Wilcox) that sold equipment to broadcasting. So that part is "for profit." But the major broadcasters in Europe are "public service broadcasters," many (as with the BBC) supported by the public (payment of a licence fee), not taking advertising and not specifically profit-oriented.[1]

While MXF is a heavily broadcaster-influenced standard, it was, as noted earlier, developed with considerable input from nonprofit broadcasters in a standards-making body (SMPTE) that is not profit-seeking, even though it charges for its standards to fund its work. The Library of Congress's work on the AS-07 specification, too, is done under the auspices of both FADGI, a collaborative body comprised of representatives from twenty-one American federal agencies, and standards-making bodies like SMPTE and AMWA, both of which have significant representation by businesses that are motivated mostly by the potential maximization of profits. For-profit entities have a very clear potential return on the investment of effort in standards like MXF and specifications like AS-07: the profits they could accrue from making and selling tools that are compliant with those technologies.[2] Furthermore, collaboration on these standards and specifications can give businesses the ability to influence them from the ground up and tailor them to their own interests. The library's adoption of MXF and JPEG 2000 and its continual adherence to them because of digital cinema, can be seen as an echo of the original public-private partnership between the library and Hollywood that had motivated the moving image copyright deposit system's development nearly a century earlier.[3] In Europe, the PREFORMA project, which provided a basis for the CELLAR standardization efforts, also took the form of a research-and-development tender pooled between numerous state and nonprofit institutions in collaboration with commercial businesses and small enterprises.

Such private-public amalgams, Nanna Bonde Thylstrup has shown, are common in many contemporary projects involving cultural heritage, such as mass digitization initiatives, because the work is costly and complicated. Thylstrup describes such partnerships as "infrapolitical," because they "hover at a frequency that lies below and beyond formal sovereign state apparatus, organized, as they are, around local—and often private or privatized—material and social infrastructures."[4] This interleaving of different actors in different scopes at different scales has intrigued standardization scholars, too,[5] some of whom have called for more scrutiny of these hybrid relations.[6] In the interest of advancing some of these concerns, the following sections contribute two examples of "infrapolitical" standards partnerships, situated against a larger backdrop of ongoing video format development as well as past communication protocol standardization.

LARGE TECH COMPANIES GET INVOLVED IN OPEN-SOURCE STANDARDS

More surprising than huge digitization initiatives is the nature and composition of the CELLAR group which, as we have discussed, is firmly grounded in the ethos of open source. CELLAR is not only comprised of representatives of nonprofit entities. Google employees are also working on the CELLAR group, and they have added material to the Matroska specification that has no direct preservation-related import. For example, as Dave Rice disclosed to us, Google wants to see support for 3D and virtual-reality video streams which could be useful for their enterprise but have no immediate utility for archives and libraries. According to Reto Kromer, there is also a continual push by Google-employed participants to keep the nascent CELLAR Matroska specification backwards-compatible with WebM video, a royalty-free container format derived from Matroska that undergirds much of YouTube.[7]

On2, the company that developed the VP8 video compression technology that forms a core part of WebM, was acquired by Google in 2010. Google immediately started transcoding the YouTube video catalog to WebM, meaning that Matroska became the container for the world's largest online digital library.[8] Steve Lhomme, Matroska's inventor, considers this a win for the format and its legitimacy. He told us he finds it amusing that his daughter, like most of the billions of YouTube users, interacts with his invention without even knowing it. Once again, Matroska proved itself to be an effective boundary object by straddling the worlds of open-source and for-profit software while serving both. However, Google's involvement in CELLAR has never been strongly promoted on the working group's website, so it may not be apparent to all CELLAR participants nor all adopters of its standards. This is not unusual, given that IETF members formally participate not as representatives of corporations but as individuals, but some scholars nonetheless describe the majority of IETF engineers as "contextually situated political brokers for their employers' interests" and a convenient method for companies to avoid corporate accountability.[9] Google's investment in the WebM backbone of YouTube and its presence in the CELLAR group's work have caused a major shift in the landscape in which the players operate. It quickly became apparent that Google's involvement, together with that of consultants like Jean-Christophe Kummer and businesses like Scenesavers, in the CELLAR work breaks down any seeming dichotomy of open-source/nonprofit and industry/for-profit. Furthermore, as Peter Bubestinger-Steindl pointed out at the second No Time To Wait conference in Vienna, open source does not necessarily mean a lack of profit.[10]

But in this way, the residue of commercial interests remains embedded in the technology, allowing us to read standards documents as traces of bygone media futures. It is useful to recall that MPEG-4 also standardized not only video compression for natural recordings but devised an entire ontology of what the standard specification calls "media objects": topological procedures for manipulating, animating, and compressing synthetic matter, such as animated meshes and textures or computer-generated face and body models. MPEG-4 thus envisaged moving image compression in tandem with interactive applications and standardized methods that could compress moving images by transferring only the instructions on how to animate them without transporting the actual animated elements. Media-historically, this is an exciting thought because it repositions the receiving machine from a deterministic decoder into an active animator that *performs* the image from a set of instructions that can be interpreted and realized in multiple ways. These MPEG-4 affordances never found much practical adoption. Yet encoded within them is a vision of what the future of moving images was expected to become at some point in the past. In the same way, provisions for 3D cinema and virtual reality in Matroska might never see the light of day in an archival setting (nor elsewhere) but Google's strong influence will ensure that CELLAR's standard carries with it at least some of the politics of that business.

While, just a few years ago, *open source* often also implicitly meant *nonprofit*, that situation is changing rapidly as large technology companies become increasingly indistinguishable from media companies. The latest generations of competing video format standards, particularly the "format war" between AV1 and VVC, are especially instructive in this regard.[11] Both of these recent compression formats are geared toward ultra-high-resolution video and directed at the end of the commercial media delivery chain, where video encounters viewers. Both formats achieve higher rates of compression (that is: smaller file sizes and better image quality). They both face off in opposite corners of the standardization arena: VVC—"Versatile Video Coding," standardized in 2020, is the youngest sibling of MPEG-2, AVC, and HEVC, the compression codecs that underlie DVD, Blu-ray, and Ultra HD Blu-ray Disc formats, respectively. All these standards carry a seal of approval from the International Standards Organization and the International Telecommunication Union. AV1, on the other hand, is an open-source standard developed by the Alliance for Open Media, an industry consortium whose founding members include, among other tech giants, Amazon, Netflix, and Google, as well as the nonprofit organization Mozilla. The primary goal of this initiative is to develop a royalty-free video compression format, driven by the tech companies' aversion to the infamously arcane and costly patent encumbrances of the earlier ISO and ITU standards family.

In the previous chapter, we have seen how audiovisual archives tend to select video preservation formats that agree with their vision of the future of media distribution. We can partly extrapolate this to the larger media and technology sectors, too. The unlikely alliance of nonprofit software foundations, smartphone and hardware manufacturers, and (under normal circumstances, rivalrous) tech and streaming companies—all banding together to develop an open-source video format outside of the regular standards trajectory—gestures toward the future of technological standardization: One in which the pedigree and recognition of the great international standard-setting bodies of the twentieth century is perhaps not yet dwindling, but, at the very least, greatly contested. New communities of practice—several distinct communities, in fact—are taking up the task of making standards and saying: We are playing by ourselves.

Netflix began streaming videos in AV1 in February 2020; YouTube followed a few weeks later, along with other large streaming services like iQIYI, Huya, and Bilibili in China, and Instagram in 2023. For AV1 to get widely adopted, these streaming providers also need to convince television, phone, and gaming console manufacturers that the broader video possibilities are worth the cost of new hardware. Electronics manufacturers (but also telecom companies, set-top box integrators, browser developers, and other more obscure actors in this complex landscape) then rely on chip designers and foundries to develop the new types of processing hardware required. The actors within this labyrinthine value chain are slowly being forced to choose and align with one of the format standards, or, like the European broadcast television standardization body DVB, equivocate and develop specifications that try to play all sides of the game to accommodate the new realities of a multi-standard world.[12]

Clearly, this book cannot be simply the story of profit versus nonprofit groups with respect to standards. While standards-making is always an exercise characterized by the push and pull of competing ethics and rationalities, efforts like CELLAR and AS-07 can be read as sites of alignment between public and private entities. Indiana University's Media Digitization and Preservation Initiative, already briefly discussed in the previous chapter, is a similar zone of convergence, made more notable for the fact that it represents a synergistic endeavor that results in the preservation of information materials that are, for the most part, owned by the state.

INDIANA UNIVERSITY'S MEDIA DIGITIZATION AND PRESERVATION INITIATIVE

According to its website, Indiana University's Media Digitization and Preservation Initiative (MDPI) was

> a six-year journey to complete the ambitious challenge of digitally preserving significant audio, video, and film assets and making them available to scholars all over the world. It became the largest university-led

> media preservation project in the United States and one of the largest in the world. By all accounts, the initiative concluded as a success with more than 350,000 items digitized and stored for posterity.[13]

The MDPI has been housed in a large facility off-campus, and people staffing the project were either employed by Indiana University or Memnon/ Sony (Fig. 4.1). Jimi found the cultures between Memnon and IU to be overlapping but also very distinct, corresponding to different material needs of the media at hand. Memnon tools and staff were mostly concerned with the high-throughput digitization of massive quantities of cassette-based video. These large, open workspaces involved multiple playback decks operating continuously throughout the week with large monitors that display many video streams simultaneously, allowing one or two staff members to oversee the digitization of many tapes at once. This system, with its very high videotape-to-staff ratio, leveraged Memnon's resources to digitize enormous quantities of materials that need minimal supervision,

FIG. 4.1. A literal black box of format standards: An LTO tape robot at Indiana University. (Photograph by Jimi Jones.)

allowing for highly efficient digitization of materials that have, on average, very few playback issues.

The high-volume digitization work contrasts and complements the work that Indiana University staff were doing in other workspaces in the same building. Indiana University staff were mostly performing digitization and conservation work on single items that require close attention and inspection. Items that are exhibiting playback issues or common breakdown syndromes like binder hydrolysis require more cautious and time-consuming handling and playback than the Memnon systems could efficiently handle, so they were set aside for the Indiana specialists. Nonetheless, the workspaces for the Indiana staff were quite different from the Memnon environments: where Memnon's mass digitization spaces are open and heavily staffed, the Indiana spaces are rooms consciously designed to allow a handful of staff to work in low-light, low-noise conditions allowing for critical listening and viewing.

After his tour of the facility, Jimi performed a group interview with several Indiana University and Memnon employees. Throughout the interview, the participants repeatedly made clear that Memnon is all about efficiency and moving quickly, while the Indiana University staff handle the objects that fall outside of the parameters of the Memnon workflow and require closer attention. Memnon and Indiana University were working within the MDPI in a way that creates and adheres to very well-defined roles that leverage the various strengths of a large for-profit technological innovator and a large nonprofit, state-funded academic institution.

Memnon, because of its size and its identity as a Sony subsidiary, has wide latitude with respect to quickly hiring staff and providing innovative technology for the playback and digitization of legacy materials. (The fact that many of the playback decks that had been pressed into service at the MDPI laboratory were, in fact, made by Sony in decades past, makes it clear that Memnon may be in a particularly advantageous position with respect to procuring hard-to-find parts for these machines' upkeep.) Indiana, on the other hand, can leverage its specialists, who have training that can take decades to acquire as well as equipment that staff has customized over decades of use, to the per-item, intensive repair, playback, and digitization of at-risk or particularly valuable items. The flow of analog and digital objects through the MDPI project, and especially the ongoing hand-off of materials from Memnon staff to Indiana University staff for their assessment, conservation, and digitization, represents a close form of collaboration between for-profit and nonprofit entities. However, as our interviewees continually asserted, there is more to this partnership than just workflows. It also represents an alliance between ethics, too. Where

Indiana University represents, to many of the MDPI workers, a slower-paced, more deliberative value system that prizes an extreme attention to detail in singular objects, the Memnon/Sony worldview holds as its highest regard the fast, efficient digitization of high quantities of material. This public-private collaboration at Indiana University shows that FFV1 and Matroska could have a bright future in higher education repositories.

The Constraints of Digital Preservation

If there is an evergreen tune in the professional literature of archives and libraries, it must be "the funding blues." The preservation of digital moving images and infrastructure incurs great cost in time, effort, and money. Archivists have become wary of electronic manufacturers' efforts to develop and market "forever media" that can supposedly reliably store information for decades. It has become obvious that such "permanent" technological preservation solutions are anything but impervious to the vagaries of materiality—physical problems or unexpected chemical phenomena—and, even more so, subject to quick cycles of technological obsolescence during which hardware, file formats, or playback software are quickly phased out in favor of newer alternatives.[14] There are so many hardware, software, and standards dependencies involved in digital preservation that the endeavor can be very challenging. Ensuring the sustainability of digital moving images requires not just technological solutions but public and private economic commitments to preservation.[15] Creating and maintaining funding streams for preservation is essential to the complicated work of keeping digital information accessible, and public-private partnerships to create digitization standards like AS-07 and PREFORMA are often critical to the larger digital preservation endeavor.

Developing best practices and standards that anticipate the needs of memory institutions is critical to responding to the growing need for planned preservation solutions,[16] and is a core recommendation of a national-level American digital preservation report (in which the Library of Congress was a participant) from 2015.[17] However, the application of these best practices often clashes with the realities of archival work and prevents archivists from using lossless standards.

TIME

A key constraint in the preservation of analog video recordings is time, or the lack of it. Videotape degrades rapidly, playback machines break down and no analog video players are currently being supported by their developers, meaning these formats have become obsolete. As one luminary in the audiovisual preservation field put it, "[a]nalog magnetic recordings are the poster children for the problems of format volatility."[18] The only uncertainty in video preservation is not *if* irreplaceable moving image recordings will be lost over time but *how many* will be lost. A commonly cited figure for the viable remaining lifespan of existing analog videotape is until roughly 2030.[19] As we have seen earlier, the creation of digital video surrogates for analog originals is currently framed as the only means by which information encoded on these vulnerable tapes can be preserved long term. Preservationists must choose from a wealth of standards for the encodings and file formats that form the digital objects they create through their digitization workflows.

The speed with which libraries and archives must move to capture magnetically recorded information before the carriers pass out of their viable playback lifespans has imposed dramatic limitations on the ability of many libraries and archives to experiment with new technologies. As Andrew DaPuzzo pointed out, the ticking clock of magnetic tape breakdown has motivated many institutions to situate themselves in what he referred to in our interview as "the 'don't give a damn' world."[20] In other words, many institutions have such a massive backlog of materials in need of digitization, coupled with insufficient staff hours to do the work, that they are disinclined to experiment with new technologies. Archivists, as Peter Bubestinger-Steindl asserted in our interview, "just need tech to work."[21] They prefer to use software, hardware, and format standards that are perceived by the preservation community as being "tried and true."

This perception often comes from the preponderance of these technologies in large institutions like the Library of Congress, television stations, Hollywood, and so on. It is often made manifest in professional email lists, organizations, and conferences where members of the for-profit and nonprofit worlds interact. Adoption of technologies tends to be self-reinforcing. The more well-adopted a standard is, the greater the perception that this standard will be a safe bet for long-term viability. This is the logic that underpins, for example, Jean-Christophe Kummer's advocacy for the AVI format. We have seen this play out in the previous chapter: The use of a standard in high-profile institutions can encourage initial adoption in other institutions. While such power to sway opinions is often initially located in large institutions, the influence becomes more distributed as adoption broadens.

The wide array of video formats can also work as a limitation for archival institutions. Magnetic videotape was designed for the logic of just-in-time image capture in televisual broadcasts and evolved into home video in the last three decades of the twentieth century. Neither of these environments tends to have a direct mandate to preserve their productions for the greater historical record—the first international, binding legal deposit requirements for television programs only entered into force in 2014. Furthermore, the succession of magnetic video media has moved at high speed as image capture and playback technologies have improved, and that has led to a wealth of incompatible formats.[22] In the words of Bijker and Pinch, "the developmental process of a technological artefact is described as an alternation of variation and selection."[23] If we understand "magnetic analog video" as a coherent technology, the ongoing success and failure cycles of formats like U-Matic, Betamax, and VHS are a perfect example of a broad, oceanic technological state with periodic and temporary swells of successful format standards and frequent whirlpools of unsuccessful ones. These failure cycles put at a disadvantage any institution not well-funded enough to collect playback equipment for the entire array of formats, particularly those that, because of market forces, did not have the benefit of being inexpensive and widely available. A further complication is that archives have little or no influence over how the materials in their collections are produced. They must be prepared to accept and preserve a wide variety of materials often not produced to any particular specifications.[24]

As Dave Rice and James Snyder both put it, audiovisual preservation standards have the most utility when donors and repositories agree on using them. This need for conformance to donor specifications often creates easy symmetries: It works for many American and European broadcast archives, where film and TV producers deliver material in the same format that they request it in when they wish to reuse it. However, the resource constraints of smaller institutions and the confusing cornucopia of formats can complicate this balance between donors and collectors. Some institutions have other constraints on which format standards they can use. For example, Lars Gaustad noted in our interview that the National Library of Norway's preservation policy postulates adherence to international policies with respect to standards and digitization. This means they must look to projects like PREFORMA for which technologies to implement. The City of Vancouver Archives, as we have seen, too, must follow state-mandated constraints to use, wherever possible, open-source technologies.

Not all institutions willingly conform to industry standards. Josh Yocum made this clear in our interview as he argued that archives should make their own decisions, informed by local practice, and, furthermore, he explained that the fact that MXF is good for broadcasters is not a selling point for him.

> [W]ith MXF, it seems like they're always talking about how it can hold so many timecodes, which, for us, that's not super important. But if you're in a broadcasting environment, that seems like something you would care about a lot more. So, a lot of the kind of features that I feel like get advertised as what makes MXF a good format [aren't] really something that we—as an archive that has video stuff [that] isn't professionally produced [. . .]—have much interest in. "How many different timecodes can this format carry?" or, "can it scale to lossless format for broadcasting easily?"[25]

For some institutions, the confluence of these circumstances results in a high-pressure, time-sensitive environment. When we spoke to Bright Joshua, Chief Conservator at the National Records and Archives Services of Malawi, he described to us an ongoing 16mm film digitization project. The digitization of these films (newsreels and documentaries of importance to Malawi's history reaching back to the early 1950s) is both chemically and administratively pressing: A part of the repository sits in the hot climate in the shadow of Zomba mountain and shortly before our conversation, the archive's outdoor air conditioning unit got stolen, and the initiative is financed by a Japanese charity on a three-year term. The project must be completed in this time and experimenting with different formats is a luxury the archive cannot afford. Unlike digitization in many places in the Northern Hemisphere, there is no scanner. The films are projected on a screen and filmed with a camera, manually color and sound corrected and checked by archive staff, and ultimately stored as high-quality but lossy MP4 files.[26] In many respects, this preservation routine does not follow the idea of what constitutes "best practice" for lossless long-term preservation according to the guidelines of international organizations. But it works. It delivers files in a ubiquitous format that the archive's users (often historians and other researchers) can use without any fuss. And that is the point. In an environment where actions have to be taken quickly, where there is external time pressure to complete the digitization, where data storage capacity is precious, where playback equipment is often unavailable to begin with,[27] where the environment adds an additional level of difficulty, and where even basic preservation infrastructure is at risk of theft or vandalism, a complex lossless workflow would not necessarily result in more favorable conditions. The infrastructure needed to implement lossless formats is simply out of reach here, and there are more acute matters to attend to.

This single digitization initiative is a microcosm of material, environmental, infrastructural, societal, political, and technological problems—and of creative solutions that make do. The situation at the Archives

Services of Malawi, which mirrors that of many audiovisual archives in Sub-Saharan Africa and elsewhere,[28] also lays bare the vast technopolitical differentials of global audiovisual preservation—but more on that in a moment.

SPACE, COST, AND ACCESS TO EXPERTISE AND DOCUMENTATION

In a world where standards are being developed online through remote conference calls and mailing lists, it may seem like geographical distance is irrelevant to standardization and standards adoption. Our interviews show that this is not the case. Spatial proximity is essential both to making an archive work and to the standards adopted within it. Inter-institutional relationships—which are also social relationships—are extremely important to how format standards propagate, and we have seen many cases in which these relationships depend on physical closeness. This plays out in public-private partnerships like that between Indiana University and Memnon, whose joint digitization project is housed, for good reason, in a single building. But it is also observable in the social structures of the film pipeline in a compact country like the Netherlands, where the local film industry is dependent on the national Film Fund, whose funding is preconditioned on the deposition of mandatory preservation copies, but which negotiates the format of those copies with the national film archive, which then individually negotiates workflows with practically all post-production studios in the country, which then tell filmmakers what they need to deliver to the archive. It is this kind of circular system of collaboration, equally buttressed by personal working relationships, logistical infrastructure, and geographical proximity that results in effective preservation. At the National Archives of Malawi, the archivists share a building with the censorship board, which employs film reviewers who grade films meant for public exhibition. When the reviewers have a spare moment, the archivists ask them to help at the adjacent archive. Like all digital technologies, video formats seem seductively placeless, but their location in physical space matters a great deal.

Another potential constraint that limits archivists' abilities to work with lossless formats is access to expertise and guidance that can help support preservation and digitization activities, particularly when problems need to be assessed and solved. As Dave Rice put it, if an archive chooses JPEG 2000, it automatically aligns with the professional cinema and broadcast industries. However, if most of the archive's peers and colleagues were part of the FFmpeg and open-source world, it might make access to expertise regarding MXF or JPEG 2000 harder than it would for FFV1 or Matroska. Being able to look up, open, and read the standards documentation itself is a critical tool for the creation and playback of digital video files. In our conversations with implementers of Matroska and FFV1, a common theme was the cost of access to the standards documents themselves. MXF has documentation that is locked behind a paywall. That can make it unappealing to some, as Jean-Christophe Kummer noted. To see the "recipe" for how MXF files are constructed, one must pay high fees to access those documents. This can slow the development of DIY tools for creating and decoding these kinds of files.

Matroska and FFV1, in contrast, both have openly accessible documentation about how to use them. Repeatedly in our interviews, the costs of standards documentation were mentioned as a barrier to access for many institutions, particularly those of modest means. In fact, the cost of SMPTE standards was a key factor in how the IETF/CELLAR work has played out, according to Ashley Blewer. Kate Murray noted that FADGI understands the problematic nature of paywalls for many institutions, so she hopes to make AS-07 a free and accessible SMPTE standard.[29] There is a paradoxical logic at work here: The constraints that smaller institutions operate under have made the adoption of costly standards difficult, but those same constraints have also set in motion the development and adoption of free and open standards like those being produced by CELLAR.

But while cost and access to technology and knowledge are integral to an archive's choice of format, it would be wrong to assume that the adoption of lossless standards corresponds with access to financing only. As we have seen, some infrastructurally robust archives in Europe, like the Netherlands Institute for Sound & Vision, use lossy compression in their video collections because their users want recordings in that format. The archive's clients trump notions like losslessness. Ultimately, archives will choose format standards that are meaningful to their users, even if they are of "sub-standard"—or, in this case, lossy—quality. For national archives in any country to adopt lossless, open-source formats like FFV1 or Matroska on a large scale, their film and broadcast industries would have to switch to those formats, too. Contrarily, some small and cash-strapped institutions end up adopting lossless formats because it helps them keep storage costs down.

Sustaining all these negotiations and workflows are trust relationships. What stood out to us across all our interviews is the tension between each archive's desire and need to be technologically independent and, simultaneously, the ubiquitous and profound dependence on the social structures that support the archival assemblage. These relationships can manifest on a regulatory level as laws and official policies, but, even then, are ultimately about people trusting each other, cooperating, and working together. One of the most exemplary and compelling cases is that of the Nigerian Film Corporation (NFC). Justina Akporherhe, film archivist at the NFC, explained to us how her archive department benefits from the technological autonomy of having its own scanner (which, in the absence of functional projectors, becomes an instrument of access, not just of digitization) and how they are simultaneously slowed down by not having their own editing suite and having to rely on the Corporation's production department to play out Betacam tapes. At the same time, she also described a huge range of personal and inter-institutional relationships that they have had to build up and maintain in order for the archive to function on the most elementary level: with the Censors Board (which keeps copies of films and might thus be useful in getting them preserved), with various film festivals, with private German film labs and digitization companies, and especially with the local filmmaking community which, as Akporherhe observed, either distrusts or does not understand the archive, and so impedes its efforts to meaningfully conduct its work. It is these networks of trust that frequently first need to be mended before a healthy environment emerges for archivists to work in. Trust, in its many forms, powers preservation: An archivist's trust in the machines she works with, her colleagues, her institution, the industry, and the standard-making organization.

The Gendered, Colonial Technopolitics of Archival Standards

"Whoever sets the standard has the power," media theorists Ned Rossiter and Geert Lovink pithily observe.[30] All of the preceding chapters have portrayed standardization as a delicate political balancing act, in the sense that it ultimately revolves around access and the governance and allocation of finite resources. But there is a more fundamental ethical problem at play here, too; one which this book cannot explore with sufficient scrutiny, yet without which any attempt to understand the politics of standards would be no more than a paltry caricature: the nature of standardization as gendered work, and that of standards as mechanisms of neo-colonial power.

When Ashley Blewer spoke to us about leaving the CELLAR group, she recalled that

> I essentially left the initiative largely around financial reasons. I couldn't sustain donating time and effort without compensation, especially when there was very little overlap with my day job. There was also a shift in

> CELLAR management at some point, and I felt like they were quite dismissive of me and my work, so maybe I would have stuck around if I felt like my opinions and efforts were valid, instead of just being tossed out. I'm not saying anyone was rude or harassing, but I did feel generally unwelcome.
>
> Except for someone who was only on the project for the first 2–3 months (and maybe a rare drop-in), I think I was the only woman on the project for the entire time. Some people are better than others around not discriminating against what I say based on my name/presentation. Yeah, so, I guess I say that I mostly stopped being involved for financial reasons, but I would have been keen to stick around and volunteer if the atmosphere was more welcoming.[31]

This was the only time that gender explicitly came up in our interviews—but then again, we had not designed our interviews to elicit insights about gender. Blewer's firsthand experiences are perhaps best interpreted in parallel with data on the representation of women in technical standardization. While research on women in standard-making is extremely scarce, it is clear that women are severely underrepresented. In 2020, of all committees within the International Electrotechnical Commission, 93 percent were chaired by men. The percentage was 87 percent at the ISO.[32] At both these standardization bodies, women tend to occupy lower-ranked, administrative positions.[33] Data from German organizations that participate in standards-making show that across all industries—but also and in particular, in information and communications technology, which includes video—women have a woefully minuscule presence in the creation of national and international standards.

The economist Philipp Heß hypothesizes that the general underrepresentation of women in technical industries results in a corresponding—but even more severe—underrepresentation in standardization, as "it is much less likely for women to reach positions in which they would be candidates for becoming an organization's representative in standardization committees."[34] Women face barriers to entry into scientific and technical fields, which range from discouraging to outright hostile and even dangerous.[35] Blewer's summary of her work in CELLAR is a reminder that the video standardization community is not immune to this, and also speaks to the precarious forms of labor that standardization sometimes entails—a precarity that is also unequally distributed. While the IETF is praised and respected for the openness of its standardization and decision-making process, more recent scientific literature has also been critical of its organizational culture. For example, the governance and standardization historians JoAnne Yates and Craig Murphy have noted that IETF's idiosyncratic etiquette might be impinging on the organization's otherwise pioneering diversity efforts, recognizing that "the direct and sometimes aggressive style of IETF meetings and electronic exchanges may make it difficult for individuals from countries with a much less direct communication style or a collectivist culture to participate. Similarly, many women, even if they are from the United States or Europe, may also be put off by this style."[36] In a detailed report on the subject, Corinnne Cath concurs that the surface image of open, informal, and apolitical Internet governance and standards organizations like the IETF conceals a predominance of corporate interests and shields them from interventions by civil society.[37] This is consistent with more than a decade of research that has found that in technically-oriented Internet communities, ideals like "openness" and "freedom" often get inflected in a way that turns them into environments that are abrasive, dismissive, and unwelcoming to women.[38]

The difficulties women face in contributing to standardization work echo the lack of participation by people outside of the Global North. For example, European archivists are overrepresented in the Technical Commission of FIAF, leading to situations wherein guidelines for archives with limited resources are being developed without archivists who have the most experience working in such conditions.[39]

Present and future video format standard development is dominated almost entirely by corporations, institutions, and individuals from Europe, North America, Japan, and China. If standardization were a perfectly rational endeavor (which it is not, as we have seen) that resulted in universally usable and equitable outputs (which it does not), then this would not necessarily be an issue. But standardization is political work, and when large communities of people are excluded from contributing to it, the social legitimacy of standards at large is at risk.

In one of the seminal texts of science and technology studies, Langdon Winner argued that technologies do not enter the world as neutral, apolitical objects. They are, instead, encoded with the biases, beliefs, and politics of the people who create them.[40] Winner cites a controversial example of this phenomenon: Robert Moses's overpasses in twentieth-century New York City. According to Winner, Moses intentionally designed these overpasses to allow cars to pass beneath them but prevented buses, which were used predominantly by marginalized communities, from doing the same. Winner argues that this design encoded a vision of parts of New York that effectively blocked poor people from entering wealthier, predominantly white neighborhoods. Winner's example reveals that seemingly neutral technologies can conceal distinct biases that both mirror and shape the prevailing social order.

In 2021, we interviewed Judith Opoku-Boateng, the senior archivist in charge of the J. H. Kwabena Nketia Archives at the University of Ghana and IASA's Vice President of Communication. Through the course of our conversation, Opoku-Boateng told us about how her archive implements MP4, a format standard developed by a group composed largely of American and European audiovisual specialists. We began to wonder, along the lines of Winner's discussion of the overpasses in parts of New York City, if a video standard like MP4 can carry certain assumptions with it and, perhaps, even impose certain kinds of order upon users from cultures that had no input into that standard. Can a video format function as a colonizing technology? Some archivists certainly perceive it that way. Opoku-Boateng also described to us the reality of international funding schemes for digitization. Many of such grants and collaborations target the preservation of at-risk collections and finance or donate machines like scanners to support preservation according to international standards. Opoku-Boateng's grievance with these generally well-intentioned initiatives is that they are often saddled with conditions that disadvantage the beneficiary institutions: they frequently involve losing control or giving up ownership over the collection.[41] Similar criticisms are increasingly heard in other professional forums, too, such as the 2022 Global Audiovisual Archiving at EYE Film Institute Netherlands, where standards and their social equitability, environmental sustainability, and international accessibility were a central point of contention. As the film archivist Carolina Cappa posits, "[a]s currently conceived, the concepts of film restoration and preservation, established by FIAF and fostered by the practices of cinematheques, film festivals and streaming platforms of the Global North, perpetuate technological-commercial dependency and promote the invisibility of the counterhegemonic cinemas and cinemas of the South."[42] Tre Berney, the Head of Digitization and Conservation at Cornell University and President of IASA, also confirmed in our interview that "[f]rom an IASA standpoint—we have heard Western technological standards described as 'oppressive.'"[43] Lossless video compression formats—the current acme of audiovisual preservation—are often unsustainable and unobtainable in many local contexts around the world.

There is no space in our book to address further issues of environmental sustainability or power consumption of digital preservation technologies, but the potency of increasingly untenable visions of "image quality" that fuel tech marketing and production practices in the media industry and ooze down into preservation has environmental ramifications, too. Despite its origins in prudent resource management, notions like "losslessness" that underlie video format standards can sometimes morph into an obsession with image resolution, detail, bit depth, dynamic resolution, light frequency spectra, and so on. The technical requirements are sometimes wildly out of proportion to their social and environmental costs, such as the amount of electricity and carbon emissions needed to produce and maintain files at such quality.[44] Such demanding standards and technologies often only serve to place preservation guidelines further out of reach of many archives and exacerbate global media inequities: a small selection of canonized films receives exclusive care and is repeatedly re-scanned and restored in exponentially growing resolutions, while the audiovisual heritage of entire countries is at risk of disappearing.

Benoît Turquety maintains that standards have "structured the global mediascape into separate networks whose frontiers further the colonial history deep into our supposed free-trade world."[45] Turquety powerfully shows this on the example of various film and video formats that have historically served as instruments of both colonial domination and resistance in several countries on the African continent, but also the much more recent digital cinema standard that effectively demoted nearly all productions in Nigeria, a global film powerhouse, into the low status of "video" rather than the culturally respectable status of "cinema." Scholars have been ambivalent about this boundary between media and formats. While some, like media scholar Ikechukwu Obiaya, recognize in Nigerian video production an ongoing liberation from the remnants of colonial filmmaking structures,[46] others consider the disappearance of a bona fide *film* industry a "tragedy."[47]

And thus, while the history of MXF, JPEG 2000, Matroska, and FFV1's standardization that we have told in this book is intricate enough on its own, our story is but a thin sliver of the complicated political position of archival format standards within global media flows. Their future use and utility in South American, Asian, and African libraries or archives will be an exciting area of future analysis. While lossless formats are, at this point, still mainly a luxury technology affordable to a growing number of archives in the Global North, their adoption among archives does not neatly map onto any easily readable cartography of global preservation practice.

In Conclusion

As technologies evolve, there is no set, natural rule saying which standards will win out over others in any given context.[48] FFV1, Matroska, MXF, and JPEG 2000 have all evolved outside of their designers' intentions and found themselves pressed into service in ways that had not been anticipated by their developers. CELLAR's work to standardize Matroska and

FFV1 will help to stabilize those technologies, as will the spread of their adoption. As we continue to move into a world where content is delivered over computers as opposed to on television and movie theater screens, FFV1 and Matroska will likely continue to have to appeal to institutions that have little or no association with those traditional avenues of dissemination. MXF and JPEG 2000, because of the difficulties many of our interviewees have expressed with respect to their implementation, are likely to continue to be preferred in institutions with strong ties to cinema and television.

The political, economic, and social forces involved in standards-making by SMPTE, the ISO, CELLAR, and the Library of Congress affect the ability of the four standards to become boundary objects that are (or are not) adopted in digital cinema, broadcasting, libraries, and archives. Standards do not emerge fully formed in any profession. They come into being from negotiations and compromises between actors. This is particularly true of audiovisual preservation, a field whose institutions are, globally speaking, structurally understaffed and underfinanced. The film and video archivist is almost always forced to be a jack-of-all-trades, improvising workflows and experimenting with machines, in constant tension between automated tools and their own sensorium. Every archive is a dynamic, messy place, even if it looks highly organized on the surface. Formats will be just as plentiful as opinions on how things should be done. Any ethos regarding preservation might clash with what money allows, what clients demand, what labs can deliver, what policy dictates, what the certifying authority wants to see, what the archivists can do, what the computers can handle, and even what the building can physically support.

Examples from our interviews have shown that even traditional institutional archives that appear stable in their internal structure are always exposed to multiple vectors of change simultaneously. The purchase of a machine can force an institution to jump into lockstep with the interests of an industry it might not have anticipated. The phasing out of celluloid productions can force the integration of a production film lab into an archive, disrupting existing patterns of practice and resulting in a complete reformatting of the archive's social structure. The appointment of a managing director who is favorably inclined toward institutional mergers can profoundly change established collection policies practically overnight, rupturing fragile equilibria and even resulting in a re-manualization of previously automated work. A simple conference presentation can cause an earthquake in the field and cast it into a decade-long identity crisis. Format standards, an unassuming technology, often record and reflect these processes in the background. Making such compromises and negotiations visible to the preservation community can help deepen the understanding of why the standards we use function as they do and why they get adopted in various institutions.

The joint AMIA/IASA conference of 2010 was, unbeknownst to many participants, a turning point in the preservation of analog and digital video. Representatives of the Library of Congress presented their work developing a specification with the Advanced Media Workflow Association for MXF and JPEG 2000. Meanwhile, the Austrian Mediathek presented about their work with FFV1. The two presentations can be viewed as representing two distinct ethics regarding standards adoption. Where the Library of Congress represents a more traditional deference to established standards bodies like SMPTE and the ISO, the Mediathek's implementation of FFV1 represents a more pragmatic, "do-it-yourself" attitude. These choices are also informed by the power dynamics within which these two entities are enmeshed. The Library of Congress, with its close ties to Hollywood, television, and its situatedness within the American copyright regime, must show preference for standards employed by large industry producers. The Austrian Mediathek, because it is not so tied down—and because it is not as well-funded—has a combination of latitude and constraints with respect to how it can and must spend its preservation resources. While not intended as a shot across the bow of "traditional" standards-based digitization, the Mediathek's contribution represented an aperture toward open-source preservation technologies.

In the decade and a half since the AMIA/IASA conference, more institutions, large and small, have been taking up the open-source cause. This is, in part, fueled by a growing availability and adoption of freeware tools for video editing, transcoding, and digitization in libraries and archives. This adoption is often motivated by pragmatic concerns related to limits on funding, staffing, and infrastructure; however, there is also a change in attitude among some librarians and archivists, characterized by a greater willingness to work at the command line and build and retrofit tools rather than rely on technologies that were not designed for archives and libraries. This attitude, for some, brings with it a strong and explicit ideological allegiance to open-source technologies.

In the cases of MXF and JPEG 2000, their evolution was linear because of the input that the film, video, and broadcasting worlds had in their designs and expansions. This input, as well as the standards' adoption in cinema and television production, meant that MXF and JPEG 2000 had a relatively easy fit with institutions like the Library of Congress and the BBC. The Library of Congress has been working on AS-07 for a decade. The library's reliance on MXF and JPEG 2000 does not just represent an alliance with cinema producers and television broadcasters. AS-07 and the standards that underlie it also represent a traditional, top-down paradigm in content distribution characterized by the strong influence media producers can exert over the means of their content distribution.

Matroska and FFV1 have had less straightforward transitions into memory institutions. Matroska, like the MP3 a decade before it, had strong associations with online piracy and illegal dissemination of audiovisual content in its first decade. Its utility in libraries and archives is one of the reasons that it has been embraced by "legitimate" actors including, in the case of the Indiana University and Memnon collaboration, one of its early detractors, Sony. This issue of piracy is, perhaps, a potential avenue for future research.

FFV1 was built as an informal experiment and lay buried within a software tool library until an archivist who was unsatisfied with "off the shelf" technologies dusted it off and gave it a "what the hell!" attempt at implementation. The growing underground experimentation with FFV1 and Matroska in archives had the feel of the subaltern for several years. The highly critical and hostile reception from many of the established preservation sages scared early adopters from going too public with their work. It was not until the CELLAR working group began their standardization that these experiments lost their "Frankenstein's unhallowed arts" patina and emerged as real, viable attempts to adapt these technologies to high-quality preservation practice. Indiana University's explicit embrace of Matroska and FFV1 has, for many institutions, sealed the deal and encouraged them to also take up the open-source cause. In the span of several years, multiple archives began using these formats and we are now beginning to see small indications of a large tectonic shift: In 2020, the Flemish Institute for Archives announced it would convert its MXF/JPEG 2000 files, a format it had adopted in 2013, to MKV/FFV1. At the end of 2023, the Library of Congress recognized MKV/FFV1 as a "preferred format." In May 2024, the popular open-source video encoding tool Handbrake added a "Preservation FFV1" export preset, which greatly simplifies the creation of such files.

Many archives belong to what is called organized civil society—those communities of practice outside of the direct market and state mechanisms that, in the words of media and standardization scholars Stefania Milan and Niels ten Oever, have "the weakest standing (and the least power) in the standard-making process."[49] Though they may not have nearly the same reach, in the modern world of streaming and downloadable content, audiovisual institutions are beginning to accumulate an agency with respect to production and dissemination that is emerging alongside the more powerful media and media tech industries. This shift, combined with the potentially ongoing need of film producers and broadcasters for archival materials in their productions, could result in FFV1 and Matroska insinuating themselves even into realms that currently are, at best, ambivalent to them.

Broadcasters, cinema producers, online video pirates, archivists, and librarians are all using MXF, JPEG 2000, Matroska, and FFV1. The technical affordances of these four format standards allow these communities to do their work, but they signify different things to the users who implement them. For example, users at the Library of Congress may view MXF and JPEG 2000 as important tools for aligning the library with television broadcasters, cinema producers, and the American copyright regime. Users of Matroska and FFV1 in a library like Indiana University's may view those tools as important signifiers of a break with large technology producers like Apple and standards bodies like SMPTE. Users of Matroska online may view that technology as a useful way to encapsulate bitstreams ripped from a Blu-ray or DVD and disseminate it online in flagrant violation of copyright policies. These technologies and their technical affordances can help them move between types of users and have meanings for them that are sometimes complementary and sometimes competing.

While the open-source community, as exemplified by the CELLAR working group, has been working to standardize encodings and formats for the cultural heritage community over the past few years, many of the institutions that would best benefit from these tools may still be unaware of them. Therefore, understanding how standards like MXF, JPEG 2000, Matroska, and FFV1 are developed is critical to preservation because there are no widely agreed-upon target digital video standards in libraries and archives. Understanding the history, development, and evolution of these standards is critical to the decision-making involved in creating sustainable digital materials and preserving audiovisual heritage worldwide.

It is too soon to know how widely AS-07 and the CELLAR standards will be adopted because the specifications have not been released to the public in their final forms as of this writing. Once they are finalized, there will be considerable promotion on professional communications channels by both of these development groups. This promotion will likely encourage experimentation with, and adoption of, these standards. In this book, we have told the story of how these format standards came to the preservation community. It is our hope that other scholars and practitioners will continue to study the development and implementation of these technologies in cultural memory contexts and broaden the analysis of their adoption to include other professions and cultures.

Notes

Introduction

1. Lewetz, interview.
2. Lewetz, interview.
3. Lewetz, interview.
4. Jones and Palmer, "A Record for Your Television."
5. Jakobs, *Standardisation Processes in IT*.
6. Fuller, *Media Ecologies*, 96.
7. Gooday and Sumner, *By Whose Standards? Standardization, Stability and Uniformity in the History of Information and Electrical Technologies*.
8. Cargill, *The Social Construction of Technological Systems*.
9. Apple, "About Incompatible Media in Final Cut Pro X."
10. FFV1 is currently in consideration for Adobe's non-linear editing software Premiere Pro.
11. Timmermans and Epstein, "A World of Standards but Not a Standard World: Toward a Sociology of Standards and Standardization."; Law, "Notes on the Theory of the Actor-Network."
12. Galloway, *Protocol*.
13. Bowker and Star, "Building Information Infrastructures for Social Worlds—The Role of Classifications and Standards."
14. Blewer, interview, April 6, 2023.
15. Busch, *Standards*, 28.
16. Murphy and Yates, *The International Organization for Standardization (ISO): Global Governance Through Voluntary Consensus*.
17. Abbate, *Inventing the Internet*; Fickers, *"Politique de la grandeur" versus "Made in Germany"*; Angulo, Calzada, and Estruch, "Selection of Standards for Digital Television"; Brookey, "The Format Wars"; Emerson, "How to Read a Network, or the Internet as Unfinished Demo."

18. Turquety, *Medium, Format, Configuration.*

19. Kwak, Lee, and Fomin, "Government Coordination of Conflicting Interests in Standardisation."

20. Schivelbusch, *The Railway Journey*; Gooday and Sumner, *By Whose Standards? Standardization, Stability and Uniformity in the History of Information and Electrical Technologies.*

21. Abbate, *Inventing the Internet*; Harcourt, Christou, and Simpson, *Global Standard Setting in Internet Governance.*

22. Koretz, *Measuring Up.*

23. Berg, *Rationalizing Medical Work*; Bowker and Star, *Sorting Things Out.*

24. Quinn, *From Artefacts to Atoms*; Schlaudt and Huber, *Standardization in Measurement.*

25. Oever and Milan, "The Making of International Communication Standards."

26. Cargill, *The Social Construction of Technological Systems.*

27. de Vries, *Standardization—A Business Approach to the Role of National Standardization Organizations.*

28. Werle, "Standards in the International Telecommunications Regime," 8.

29. Arthur, "Competing Technologies, Increasing Returns, and Lock-In by Historical Events."

30. Bowker and Star, "Building Information Infrastructures for Social Worlds: The Role of Classifications and Standards."

31. Anselm Strauss, Shizuko Fagerhaugh, Barbara Suczek and Carolyn Wiener quoted in Bowker and Star.

32. Mason, "What Is Complexity Theory and What Are Its Implications for Educational Change?"

33. Timmermans and Epstein, "A World of Standards but Not a Standard World: Toward a Sociology of Standards and Standardization," 71.

34. Verbruggen, interview.

35. Ermoshina and Musiani, "'Standardising by Running Code'"; Also Oever and Milan, "The Making of International Communication Standards."

36. Pedersen, Fomin, and Vries, "Open Standards and Government Policy"; Blind and Gauch, "Trends in ICT Standards"; Brunswicker, Rodriguez, and Wareham, "Governing Standardization through Formal and Informal Standard Development Organizations."

37. Hanseth et al., "Reflexive Standardization."

38. Sterne, *MP3.*

39. Mulvin and Sterne, "Scenes from an Imaginary Country: Test Images and the American Color Television Standard—Dylan Mulvin, Jonathan Sterne, 2016," 24.

40. Sterne, *MP3.*

41. Volmar, Jancovic, and Schneider, "Format Matters: An Introduction to Format Studies."

42. Sterne, *MP3*; McGill, "Format."
43. Turquety, *Medium, Format, Configuration*, 37.
44. Wasson, "Formatting Film Studies."
45. McGill, "Format"; Fahle et al., "Medium | Format."
46. Quoted in Kuhr, "Implementation of a JPEG Decoder for a 16-Bit Microcontroller."
47. Lawrence et al., *Risk Management of Digital Information*; Fresa, Justrell, and Prandoni, "Digital Curation and Quality Standards for Memory Institutions."
48. Sterne, *MP3*, 22.
49. Volmar, "Formats as Media of Cooperation."
50. Timmermans and Epstein, "A World of Standards but Not a Standard World: Toward a Sociology of Standards and Standardization."
51. Sterne, *MP3*, 8.
52. Sterne, *MP3*, 22.
53. Belton, "Historical Paper"; Decherney, *Hollywood's Copyright Wars*.
54. Blewer, interview, April 6, 2023.
55. Lischer-Katz, "Considering JPEG2000 for Video Preservation," 1063.
56. Larkin, "The Materiality of Cinema Theaters in Northern Nigeria"; Lobato, *Shadow Economies of Cinema*; Heller, "Warum Filmgeschichte?"; Schneider, "Viewer's Digest: Small Gauge and Reduction Prints as Liminal Compression Formats."
57. Fairbairn, Pimpinelli, and Ross, *The FIAF Moving Image Cataloguing Manual*.
58. e.g., Smither, "Formats and Standards."
59. Van House and Churchill, "Technologies of Memory: Key Issues and Critical Perspectives."
60. Holdsworth, "Television Resurrections."
61. Gracy, "Moving Image Preservation and Cultural Capital."
62. Slide, *Nitrate Won't Wait: A History of Film Preservation in the United States*.
63. Wright, "Preserving Moving Pictures and Sound."
64. Ward, *A Manual of Sound Archive Administration*, 199.
65. Lesk, "Preservation of New Technology. A Report of the Technology Assessment Advisory Committee to the Commission on Preservation and Access," 13.
66. Graham, *Intellectual Preservation*; Walters, "Thinking About Archival Preservation in the '90s and Beyond."
67. Waters et al., "The Digital Preservation Consortium Mission and Goals."
68. The Science and Technology Council of the Academy of Motion Picture Arts and Sciences, "The Digital Dilemma: Strategic Issues in Archiving and Accessing Digital Motion Picture Materials."
69. Rubin, "Preserving Digital Public Television Final Report."
70. Pickren, "Political Ecologies of Electronic Waste."

71. Bradley, "Defining Digital Sustainability."

72. Fossati, *From Grain to Pixel*.

73. Robinson, "Remembering Things Differently."

74. Frick, *Saving Cinema*.

75. Kaufman, "Towards a New Audiovisual Think Tank for Audiovisual Archivists and Cultural Heritage Professionals."

76. Fleischhauer and Bradley, "IASA-TC 06 Guidelines for the Preservation of Video Recordings," B-34.

77. Bird, interview.

78. Bird, interview.

79. https://www.digitalrebellion.com/webapps/videocalc, accessed November 23, 2018.

80. Gilmour and Davila, "Lossless Video Compression for Archives: Motion JPEG2k and Other Options," 1.

81. Manovich, *The Language of New Media*; Menkman, "A Vernacular of File Formats—A Guide to Databend Compression Design"; Sterne, *MP3*; Mackenzie, "Every Thing Thinks: Sub-Representative Differences in Digital Video Codecs"; Galloway and LaRivière, "Compression in Philosophy"; Jancovic, *A Media Epigraphy*.

82. Donaldson and Yakel, "Secondary Adoption of Technology Standards"; Lischer-Katz, "Considering JPEG2000 for Video Preservation"; Lischer-Katz, "Studying the Materiality of Media Archives in the Age of Digitization."

83. Hediger, "The Original Is Always Lost"; also Fossati, *From Grain to Pixel*.

84. Lischer-Katz, "Studying the Materiality of Media Archives in the Age of Digitization."

85. Drucker, "Performative Materiality."

86. Carse, "Nature as Infrastructure."

87. Lischer-Katz, "Studying the Materiality of Media Archives in the Age of Digitization," n.p.

88. Kling, "What Is Social Informatics and Why Does It Matter?"

89. Bishop and Star, "Social Informatics of Digital Library Use and Infrastructure."

90. Kling, "What Is Social Informatics and Why Does It Matter?"

91. Meyer, *Examining the Hyphen: The Value of Social Informatics for Research and Teaching. In Social Informatics: Past, Present and Future*.

92. MacKenzie and Wajcman, *The Social Shaping of Technology*.

93. Kling, "Learning About Information Technologies and Social Change: The Contribution of Social Informatics."

94. Bishop and Star, "Social Informatics of Digital Library Use and Infrastructure."

95. Weber, "Recording on Film, Transmitting by Signals."

96. Pinch and Bijker, "The Social Construction of Facts and Artifacts: Or How the Sociology of Science and the Sociology of Technology Might Benefit Each Other."

97. Pinch and Bijker, "The Social Construction of Facts and Artifacts."

98. Rossell, "Die soziale Konstruktion früher technischer Systeme der Filmprojektion"; Weber, "Recording on Film, Transmitting by Signals."

99. Genus and Coles, "Rethinking the Multi-Level Perspective of Technological Transitions," 1438.

100. Denegri-Knott, "Sinking the Online 'Music Pirates:' Foucault, Power and Deviance on the Web."

101. Sterne, *MP3*.

102. Akrich, "The De-Scription of Technical Objects."

103. Woolgar, "Configuring the User."

104. Winner, "Do Artifacts Have Politics?"

105. Klein and Kleinman, "The Social Construction of Technology"; Prell, "Rethinking the Social Construction of Technology through 'Following the Actors.'"

106. The plethora of currents within science and technology studies has been historicized, for example, in Baron and Gomez, "The Associations between Technologies and Societies."

107. Larkin, *Signal and Noise*.

108. Campbell, *Wireless Writing in the Age of Marconi*; Mackenzie, *Wirelessness*.

109. Galloway, *Protocol*; Chun, *Control and Freedom*; van Dijck, *The Culture of Connectivity*.

110. Roth, "Looking at Shirley, the Ultimate Norm"; Mulvin and Sterne, "Scenes from an Imaginary Country: Test Images and the American Color Television Standard—Dylan Mulvin, Jonathan Sterne, 2016"; Menkman, "Behind the White Shadows of Image Processing: Shirley, Lena, Jennifer and the Angel of History"; Mulvin, *Proxies*.

111. Murphy and Yates, *The International Organization for Standardization (ISO): Global Governance Through Voluntary Consensus*; Villarreal, *International Standardization and the Agreement on Technical Barriers to Trade*; Balbi and Fickers, *History of the International Telecommunication Union*. Cf. Yates and Murphy, *Engineering Rules*, where questions of gender appear more prominently.

112. Wajcman, *Feminism Confronts Technology*; Cockburn and Ormrod, *Gender and Technology in the Making*; Calvert and Terry, *Processed Lives*; Fox, Johnson, and Rosser, *Women, Gender, and Technology*.

113. Thylstrup, *The Politics of Mass Digitization*.

114. Bischofberger, "Moving Ahead with Gender Diversity at the IEC"; Parkouda and Liao, "Understanding the Relationship Between the Gender Composition of Technical Committees and Performance."

115. Thylstrup, *The Politics of Mass Digitization*, 31.

116. Oever and Milan, "The Making of International Communication Standards."

117. Oever and Milan, "The Making of International Communication Standards," 16.

118. Holstein and Gubrium, *The Active Interview*.

119. Kvale, *InterViews*.

120. Bourdieu and Wacquant, *An Invitation to Reflexive Sociology*; Schäfer, *HabitusAnalysis 2—Praxeology and Meaning*.

121. Schmidt, "The Methodological Challenges of Practicising Praxeology," 49.

122. Ermoshina and Musiani, "'Standardising by Running Code,'" 348.

123. Devlin and Wilkinson, *The MXF Book: An Introduction to the Material eXchange Format*; Taubman and Marcellin, *JPEG 2000: Image Compression Fundamentals, Standards and Practice*.

Chapter 1. "Lossless"

1. e.g., Kennelly, *Artificial Electric Lines*.

2. Huffman, "Notes on Information-Lossless Finite-State Automata," 41.

3. Kerntopf et al., "Early Pioneers to Reversible Computation."

4. Saager and Bührer, *Die Organisierung der geistigen Arbeit durch "die Brücke," von K.W. Bührer und Ad. Saager*.

5. Holt, "Wilhelm Ostwald's 'The Bridge'."

6. Krajewski, *Restlosigkeit*.

7. Kinross, *A4 and Before*; Schmidt-Bachem, *Aus Papier*.

8. Bartlett, "Algebra, the Golden Rectangle, and Paper Sizes," 413.

9. Dunn, *Notes on the Standardization of Paper Sizes*.

10. FIAF Technical Commission, "Digitisation for Film Archives—Assorted Complications," 3.

11. Tansey, "Archival Adaptation to Climate Change."

12. Gant, interview.

13. We thank Maral Mohsenin for these concrete examples.

14. Heydegger, "Analysing the Impact of File Formats on Data Integrity."

15. I.e. Prentice and Gaustad, *IASA-TC 03*, 14.

16. See, e.g., O'Leary, "Introduction to FFV1 and Matroska for Film Scans."

17. Lischer-Katz, "Considering JPEG2000 for Video Preservation," 1057.

18. Lischer-Katz, "Considering JPEG2000 for Video Preservation," 1056.

19. Star, "The Ethnography of Infrastructure," 59.

20. Wasson, "Formatting Film Studies," 59.

21. *Free Software and Open Formats*.

22. Hayles, *How We Became Posthuman*, 13.

23. Blanchette, "A Material History of Bits." Schüttpelz, "Die medientechnische Überlegenheit des Westens. Zur Geschichte und Geographie der immutable mobiles Bruno Latours." provides a critical history of the myth of the immutability of print media.

24. Njoroge, Rath, and Sammons, "The Work of Vinyl in the Age of Digital Reproduction." Also Chun, "The Enduring Ephemeral."

25. Kirschenbaum, *Mechanisms*, 11, our emphasis.

26. Blanchette, "A Material History of Bits."

27. Smith quoted in Capell 2010, 242–43.

28. IASA Technical Committee 2014, 4. Also J. Snyder 2011.
29. Cherchi Usai, "The Demise of Digital (Print #1)"; Edmondson, *Audiovisual Archiving*.
30. Gant, interview.
31. Gant, interview.
32. Doane, "The Indexical and the Concept of Medium Specificity," 142.
33. Gant, interview.
34. Pavuza, "Migration of Archived Video Footage."
35. Rice, interview, February 21, 2018.
36. Pavuza, "Migration of Archived Video Footage."
37. Pavuza, "Migration of Archived Video Footage," 62.
38. Geimer, "Das Bild als Spur. Mutmaßung über ein untotes Paradigma."
39. Kulovits et al., "From TIFF to JPEG 2000?"
40. Gant, interview.
41. de Jong, "Sound and Vision—Digital Preservation: Policy, Standards and Procedures."
42. Verbruggen, interview.
43. Carl Fleischhauer of the Library of Congress and Jim Lindner, video engineer and entrepreneur.
44. Snyder, interview. "9/7" here refers to the Cohen-Daubechies-Feauveau 9/7 wavelet transform, a specific compression method selected for lossy JPEG 2000 (and used, among other applications, for fingerprint compression by the FBI).
45. Rice, interview, February 21, 2018.
46. Rice, interview.
47. Rice, interview.
48. Foucault, *The Archaeology of Knowledge*; Foucault, "Politics and the Study of Discourse."
49. Foucault, "Politics and the Study of Discourse," 59.
50. Lorrain, "Interview with Peter Bubestinger and Herman Lewetz."
51. O'Leary, "Introduction to FFV1 and Matroska for Film Scans," n.p.
52. Rice, interview, February 21, 2018.
53. On infrared scanning, see also Stanco, Battiato, and Gallo, *Digital Imaging for Cultural Heritage Preservation Analysis, Restoration, and Reconstruction of Ancient Artworks*.
54. Bubestinger-Steindl, "Review and Comparison of FFV1 versus Other Lossless Video Codecs for Long-Term Preservation."
55. Fairbairn, Pimpinelli, and Ross, *The FIAF Moving Image Cataloguing Manual*.
56. For example, BBC's *Ingex* and Austrian Mediathek's *DVA-Profession*.
57. Such informal knowledge practices are crucial to the adoption of a format. Compare this with the *Independent JPEG Group* which was also organized mainly via informal mailing lists and Usenet newsgroups, and was a major contributor to the later global success of the JPEG image format.

58. On this point, see also Heller and Flueckiger, "Zur Wertigkeit von Filmen. Retrodigitalisierung und Filmwissenschaft."

59. Hughes, "The Evolution of Large Technological Systems."

60. Smither, "Formats and Standards," 327.

61. On this, see especially Drucker, "Performative Materiality." And in the context of digital television, Stauff, "Das neue Fernsehen. Machteffekte einer heterogenen Kulturtechnologie."

62. Bowker and Star, *Sorting Things Out*.

63. Star, "The Structure of Ill-Structured Solutions."

64. Gasser, "The Integration of Computing and Routine Work"; Gießmann and Schabacher, "Umwege und Umnutzung oder: Was bewirkt ein »Workaround«?"

65. Sterne, "Out with the Trash: On the Future of New Media."

66. Kromhout, "Noise Resonance: Technological Sound Reproduction and the Logic of Filtering."

67. Heslop, Davis, and Wilson, "An Approach to the Preservation of Digital Records"; Keitel, "Der nestor-Leitfaden zur Digitalen Bestandserhaltung und seine Folgen für die Archive."

68. Lesk, "Preservation of New Technology. A Report of the Technology Assessment Advisory Committee to the Commission on Preservation and Access."; IASA Technical Committee, *Handling and Storage of Audio and Video Carriers (TC 05)*.

69. For an in-depth discussion of the utility and limitations of checksums in audiovisual preservation, see Rice, "Reconsidering the Checksum for Audiovisual Preservation: Detecting Digital Change in Audiovisual Data with Decoders and Checksums."

70. Blewer and Rice, "Status of CELLAR: Update from an IETF Working Group for Matroska and FFV1," 30.

71. Henderson, interview.

72. O'Leary, "Introduction to FFV1 and Matroska for Film Scans."

73. Noorlander, interview.

74. Capell, "Digitization as a Preservation Method for Damaged Acetate Negatives," 239.

75. Kokaram, "On Missing Data Treatment for Degraded Video and Film Archives."

76. Lischer-Katz, "Considering JPEG2000 for Video Preservation," 1062.

77. Jancovic, "Lossless Compression."

78. Heller and Flueckiger, "Zur Wertigkeit von Filmen. Retrodigitalisierung und Filmwissenschaft."

79. Blewer, interview, April 6, 2023.

80. Akporherhe, interview.

81. Whitehead, *Memory*.

82. Chun, "The Enduring Ephemeral," 168, our emphasis.

83. Assmann, "Canon and Archive," 106.

84. Blanchette and Johnson, "Data Retention and the Panoptic Society"; Mayer-Schönberger, *Delete*; Flower, *The Art of Forgetting*; de Klerk, "Materiality of Deletion."

85. Adorno, "The Curves of the Needle," 48.

86. Doane, "The Indexical and the Concept of Medium Specificity," 144–45.

87. See, e.g., Lischer-Katz, "Studying the Materiality of Media Archives in the Age of Digitization."

88. Doane, "The Indexical and the Concept of Medium Specificity," 143.

89. Smith and Hennessy, "Anarchival Materiality in Film Archives."

90. Thylstrup, *The Politics Of Mass Digitization*, 7.

91. This slippage that was, apropos, also at the heart of Friedrich Kittler's critique of software. Kittler, "There Is No Software."

92. Lischer-Katz, "Studying the Materiality of Media Archives in the Age of Digitization," n.p.

Chapter 2. Standards with a Capital S

1. Star and Griesemer, "Institutional Ecology, 'Translations' and Boundary Objects," 393.

2. Winner, "Do Artifacts Have Politics?"

3. Yocum, "Audio/Video Digitization Guidelines for Northwestern Library."

4. IASA Technical Committee, *Guidelines for the Preservation of Video Recordings (TC 06)*, 2019.

5. Rubin, "Preserving Digital Public Television Final Report."

6. Federal Agencies Digitization Guidelines Initiative (FADGI), "Audio-Visual Format Documentation Project: Background Paper Federal Agencies Audio-Visual Digitization Guidelines Working Group."

7. Blood, "Refining Conversion Contract Specifications: Determining Suitable Digital Video Formats for Medium-Term Storage," 26.

8. Devlin, "MXF—the Material eXchange Format," 1.

9. Rubin, "Preserving Digital Public Television Final Report."

10. Werle, "Standards in the International Telecommunications Regime," 38.

11. Bowker and Star, "Building Information Infrastructures for Social Worlds: The Role of Classifications and Standards."

12. Wright, "Preserving Moving Pictures and Sound."

13. Devlin and Wilkinson, *The MXF Book: An Introduction to the Material eXchange Format.*

14. Devlin and Wilkinson, *The MXF Book.*

15. Devlin and Wilkinson, *The MXF Book.*

16. Ferreira, "MXF—a Progress Report," 2.

17. de Vries, *Standardization—A Business Approach to the Role of National Standardization Organizations.*

18. Taubman and Marcellin, *JPEG 2000: Image Compression Fundamentals, Standards and Practice*, 5.

19. Jancovic, *A Media Epigraphy*.

20. Campbell, "The Great Video Codec Debate: JPEG 2000 and FFV1."

21. Taubman and Marcellin, *JPEG 2000: Image Compression Fundamentals, Standards and Practice*.

22. Nowak and Fößel, "Proposal of a System Architecture for Digital Film Archives Using JPEG 2000 and MXF."

23. Taubman and Marcellin, *JPEG 2000: Image Compression Fundamentals, Standards and Practice*.

24. Devlin and Wilkinson, *The MXF Book: An Introduction to the Material eXchange Format*.

25. https://lsv.uky.edu/scripts/wa.exe?A2=ind1810&L=AMIA-L&P=86917, accessed January 8, 2019.

26. Snyder, interview.

27. Sterne, *MP3*.

28. Jarczyk, Kromer, and Pfluger, "Memoriav Recommendations Digital Archiving of Film and Video: Principles and Guidance."

29. DaPuzzo, interview.

30. Snyder, interview.

31. Gaustad, interview.

32. Ferreira, "MXF—a Progress Report," 7.

33. Bowker and Star, "Building Information Infrastructures for Social Worlds: The Role of Classifications and Standards."

34. Fleischhauer, "User Needs and MXF Options: Preservation Planning and the AS-07 Specification."

35. Snyder, "JPEG2000 in Moving Image Archiving."

36. Fleischhauer, "User Needs and MXF Options: Preservation Planning and the AS-07 Specification," 5.

37. Murray, interview.

38. Federal Agencies Digital Guidelines Initiative (FADGI), "AS-07: MXF Archive and Preservation Format Application Specification."

39. Snyder, interview.

40. Murray, interview.

41. Snyder, "JPEG2000 in Moving Image Archiving."

42. Jancovic, *A Media Epigraphy*.

43. The digital cinema format DCP went through a comparable process. The Digital Cinema Initiative, the alliance of large Hollywood film studios behind the standard, initially included only two projection frame rates in the DCP specification: 24 and 48 frames per second, with no provisions for other speeds. This was a major issue for historical films, which are often projected at other frame rates. Already in 2007, the International Federation of Film Archives complained about the inadequacy of this specification, as it provides no way of showing lower historical frame rates in a digital cinema. In response, a number of European organizations teamed up to produce a new ISO standard in 2011. It included a palette of six frame rates, including

television broadcast rates. It was implemented by cinema server manufacturers fairly quickly, although many large multiplex cinemas held out on the expensive upgrades to their equipment, thus limiting the standard's utility. For digitized early films, even the new standard fails to accurately recreate the conditions of early cinema, since the issue is not to support more fixed frame rates but to allow films to be played with fluid speed so as to emulate the organic pace of, for example, hand-cranked projection. See also Jancovic, "Ghosts of the Past: Frame Rates, Cranking and Access to Early Cinema."

44. Murray, interview.

45. Murray, interview.

46. Fleischhauer, "Comparing Formats."

47. Murray, interview.

48. http://www.digitizationguidelines.gov/guidelines/MXF_app_spec.html, accessed January 24, 2019.

49. Kaghan and Bowker, "Out of Machine Age?"

50. Lorrain, "A Short Guide to Choosing a Digital Format for Video Archiving Masters."

51. Bubestinger-Steindl, "File Formats for Audiovisual Preservation: How to Choose?"

52. Blewer, interview, September 13, 2017.

53. Krabbenhoeft, interview.

54. Gaustad, interview.

55. Declercq, interview.

56. Bird, interview.

57. Gaustad, interview.

58. Lorrain, "A Short Guide to Choosing a Digital Format for Video Archiving Masters."

59. Bubestinger-Steindl, interview.

60. Mason, "What Is Complexity Theory and What Are Its Implications for Educational Change?"

61. Casey, "Encoding and Wrapper Decisions and Implementation for Video Preservation Master Files," 3.

62. Winner, "Do Artifacts Have Politics?" 134.

63. https://support.apple.com/en-us/HT209000, accessed December 17, 2018.

64. Bowker and Star, "Building Information Infrastructures for Social Worlds: The Role of Classifications and Standards."

65. MacKenzie and Wajcman, *The Social Shaping of Technology*.

Chapter 3. Wild Formats

1. Bradner, "The Internet Standards Process—Revision 3."

2. https://datatracker.ietf.org/doc/review-ietf-cellar-ffv1-03-genart-early-miller-2018-07-05/, accessed August 3, 2023.

3. Verbruggen, interview.

4. Blewer and Rice, "Status of CELLAR: Update from an IETF Working Group for Matroska and FFV1."

5. Blewer and Rice, "Status of CELLAR."

6. Lorrain, "Interview with Peter Bubestinger and Herman Lewetz." The document referred to is Lomonosov Moscow State University Graphics & Media Lab (Video Group), "Lossless Video Codecs Comparison 2007."

7. Brunsson and Jacobsson, "Following Standards," 132.

8. Niedermayer, "FFV1 Video Codec Specification."

9. FADGI (The Federal Agencies Digitization Guidelines Initiative) Audio-Visual Working Group, "FADGI"; Bubestinger, Lewetz, and Jaks, "Comparing Video Codecs and Containers for Archives"; Marsh, "The Pros and Cons of FFV1."

10. Lorrain, "Interview with Peter Bubestinger and Herman Lewetz."

11. Dave Rice and Peter Bubestinger reveal some more background information about FFV1 in this Google Groups discussion thread from 2012: https://groups.google.com/forum/?fromgroups=#!topic/archivematica/HulV96gJ0go.

12. Snyder, "JPEG2000 in Moving Image Archiving."

13. FIAT/IFTA, "Call for Presentations FIAT/IFTA World Conference Vienna 7–10 October 2015."

14. Fleischhauer, "Comparing Formats."

15. It is worth pointing out that, depending on the number of checksums generated and embedded in its bit stream, an FFV1 object could also be computationally intensive to produce.

16. IASA Technical Committee, *Guidelines for the Preservation of Video Recordings (TC 06)*, 2019.

17. Lobato, *Shadow Economies of Cinema*, 2012.

18. Blewer and Rice, "Status of CELLAR: Update from an IETF Working Group for Matroska and FFV1."

19. Yocum, "Audio/Video Digitization Guidelines for Northwestern Library."

20. Lhomme, interview.

21. Lhomme, interview.

22. Volmar, "Das Format als medienindustriell motivierte Norm: Überlegungen zu einem medienkulturwissenschaftlichen Formatbegriff." Also Sterne, *MP3*.

23. Philpott, "Changes to MKV Format Support in Premiere Pro and Adobe Media Encoder."

24. Mosco, *The Political Economy of Communication: Rethinking and Renewal*.

25. On piracy as preservation, see De Kosnik, "The Collector Is the Pirate"; De Kosnik, *Rogue Archives*; Lobato, *Shadow Economies of Cinema*.

26. Krabbenhoeft, interview.

27. https://www.loc.gov/preservation/digital/formats/fdd/fdd000342.shtml, accessed January 8, 2019.

28. McLellan, interview.

29. Casey, "Encoding and Wrapper Decisions and Implementation for Video Preservation Master Files."

30. Casey, "Encoding and Wrapper Decisions and Implementation for Video Preservation Master Files."

31. Casey, "Encoding and Wrapper Decisions and Implementation for Video Preservation Master Files," 7.

32. Blewer, interview, April 6, 2023.

33. Snyder, interview.

34. Mobley, interview.

35. Bird, interview.

36. Kromer, interview.

37. Kummer, interview.

38. IASA Technical Committee, *Handling and Storage of Audio and Video Carriers (TC 05)*.

39. Bowker, *Memory Practices in the Sciences*; Gitelman, *"Raw Data" Is an Oxymoron*.

40. Jancovic, "Lossless Compression."

41. Blewer and Rice, "Status of CELLAR: Update from an IETF Working Group for Matroska and FFV1."

42. Blewer, interview, September 13, 2017.

43. https://www.ietf.org/mailman/listinfo/cellar, accessed January 8, 2019.

44. CELLAR, "Codec Encoding for LossLess Archiving and Realtime Transmission."

45. Mosco, *The Political Economy of Communication: Rethinking and Renewal*.

46. Jenkins, *Convergence Culture*.

47. Schmidt and Werle, *Coordinating Technology*.

48. Fallon, Rice, and Blewer, "Conch—Appendix on Standardization Exercises."

49. Fallon, Rice, and Blewer, "Conch—Appendix on Standardization Exercises," 2.

50. Fallon, Rice, and Blewer, "Conch—Appendix on Standardization Exercises."

51. For more information on the way IETF operates, see Harcourt, Christou, and Simpson, *Global Standard Setting in Internet Governance*.

52. Blewer, interview, September 13, 2017.

53. Abbate, *Inventing the Internet*; Oever and Milan, "The Making of International Communication Standards."

54. Oever and Milan, "The Making of International Communication Standards."

55. Blewer and Rice, "Status of CELLAR: Update from an IETF Working Group for Matroska and FFV1."

56. Snyder, interview.

57. https://datatracker.ietf.org/doc/rfc8493/, accessed May 21, 2019.

58. DaPuzzo, interview.

59. Dunn, interview.
60. Blewer, interview, September 13, 2017.
61. Gaustad, interview.
62. Balsom, "From Singular to Plural."
63. Robinson, "Remembering Things Differently."
64. Cappa, "Sur, paredón y después. La conferencia internacional Eye 2022," 34, our translation.
65. Lewetz, interview.
66. Kummer, interview.
67. Lorrain, "Interview with Peter Bubestinger and Herman Lewetz."
68. Turkus, interview.
69. Declercq, interview.
70. Rice, interview, September 8, 2017.
71. Schupert, interview.
72. Bubestinger-Steindl, interview.
73. Blewer, interview, September 13, 2017.
74. Dix, interview, September 8, 2017.
75. Winner, "Do Artifacts Have Politics?"
76. Dix, interview, September 8, 2017.
77. Lorrain, "A Short Guide to Choosing a Digital Format for Video Archiving Masters."
78. Gaustad, interview.
79. Bubestinger-Steindl, interview.
80. It is remarkable that this seeming generational divide seems to persist over decades within standards-setting bodies. Kai Jakobson had already commented on the young average age of IETF's workgroup members back in 2000: Jakobs, *Standardisation Processes in IT*, 172.
81. Mosco, *The Political Economy of Communication: Rethinking and Renewal.*
82. Bubestinger-Steindl, interview.
83. Casey, interview.
84. Kummer, interview.
85. DaPuzzo, interview.
86. Devlin and Wilkinson, *The MXF Book: An Introduction to the Material eXchange Format.*
87. Jaks, interview.
88. https://lsv.uky.edu/scripts/wa.exe?A2=ind1810&L=AMIA-L&P=86917, accessed November 30, 2018.
89. Pinch and Bijker, "The Social Construction of Facts and Artifacts: Or How the Sociology of Science and the Sociology of Technology Might Benefit Each Other."
90. Blewer, interview, April 6, 2023.
91. Pinch, "Giving Birth to New Users: How the Mini-Moog Was Sold to Rock and Roll."
92. https://www.webmproject.org/about/, accessed January 8, 2019.

93. Suchman, Blomberg, and Trigg, "Reconstructing Technologies as Social Practice."

94. https://www.Matroska.org/technical/guides/faq/index.html, accessed April 3, 2019.

95. Bubestinger, Lewetz, and Jaks, "Comparing Video Codecs and Containers for Archives."

96. Wheeler, interview.

97. Bird, interview.

98. National Digital Stewardship Alliance, "National Agenda For Digital Stewardship."

99. Keenan, "Counter-Forensics and Photography"; Anand, "10 Thesis [*sic*] on the Archive"; De Kosnik, *Rogue Archives*; Kozel, *Openness*; Smolicki, "Para-Archives: Rethinking Personal Archiving Practices in the Times of Capture Culture."

Chapter 4. Standards at Work

1. Richard Wright, personal email communication, October 30, 2018.

2. It is worth noting that as of 2022, the Library of Congress's AS-07 document has been updated to include mapping of FFV1 to MXF: https://www.digitizationguidelines.gov/guidelines/MXF_app_spec.html, accessed September 29, 2023.

3. Frick, *Saving Cinema*.

4. Thylstrup, *The Politics of Mass Digitization*, 27.

5. Star and Lampland, "Reckoning with Standards," 15–17.

6. Oever and Milan, "The Making of International Communication Standards," 16.

7. Kromer, interview.

8. Blizzard, "Firefox, YouTube and WebM—Mozilla Hacks."

9. Cath, "Loud Men Talking Loudly," 12.

10. https://www.youtube.com/channel/UC-NF6EF-tN0S0FrJUD20-ww/videos, accessed April 4, 2019.

11. Caballar, "Battle of the Video Codecs."

12. Hunter, "DVB Selects AV1 With AVS3 And VVC As Candidate Codecs For Standardization."

13. https://mdpi.iu.edu/, accessed June 30, 2023.

14. Bradley, "Defining Digital Sustainability."

15. Blue Ribbon Task Force on Sustainable Digital Preservation and Access, "Sustainable Economics for a Digital Planet."

16. Smith, "Valuing Preservation."

17. National Digital Stewardship Alliance, "National Agenda for Digital Stewardship."

18. Fleischhauer, "Looking at Preservation from the Digital Library Perspective," 96.

19. Casey, "Why Media Preservation Can't Wait."

20. DaPuzzo, interview.
21. Bubestinger-Steindl, interview.
22. Wasser, *The Hollywood Empire and the VCR*.
23. Pinch and Bijker, "The Social Construction of Facts and Artifacts," 28.
24. Fleischhauer, "Looking at Preservation from the Digital Library Perspective."
25. Yocum, interview.
26. Joshua, interview.
27. Opoku-Boateng et al., "The J.H. Kwabena Nketia Archive at the University of Ghana- Legon."
28. Rakemane and Mosweu, "Challenges of Managing and Preserving Audio-Visual Archives in Archival Institutions in Sub Saharan Africa."
29. Murray, interview.
30. Lovink and Rossiter, "Seriality for All: The Role of Protocols and Standards in Critical Theory," 426.
31. Blewer, interview, April 6, 2023.
32. Heß, "SDG 5 and the Gender Gap in Standardization."
33. Heß, "SDG 5 and the Gender Gap in Standardization."
34. Heß, "SDG 5 and the Gender Gap in Standardization," 6.
35. Bystydzienski and Bird, *Removing Barriers*; Sugimoto and Larivière, *Equity for Women in Science*.
36. Yates and Murphy, *Engineering Rules*, 337.
37. Cath, "Loud Men Talking Loudly," 12.
38. Reagle, "'Free as in Sexist?'"; Ensmenger, "'Beards, Sandals, and Other Signs of Rugged Individualism'"; Brooke, "'Condescending, Rude, Assholes.'"
39. Blot-Wellens and Fourier, "Minimum Viable Archiving."
40. Winner, "Do Artifacts Have Politics?"
41. Opoku-Boateng, interview.
42. Cappa, "Sur, paredón y después. La conferencia internacional Eye 2022," 32, our translation.
43. Berney, interview.
44. Fossati, "For a Global Approach to Audiovisual Heritage."
45. Turquety, *Medium, Format, Configuration*, 25.
46. Obiaya, "A Break with the Past"; Petty, "Digital Video Films as 'Independent' African Cinema."
47. Olayiwola, "From Celluloid to Video."
48. Bowker and Star, "Building Information Infrastructures for Social Worlds."
49. Oever and Milan, "The Making of International Communication Standards," 2.

Works Cited

Abbate, Janet. *Inventing the Internet*. Cambridge, MA: MIT Press, 2000.

Adorno, Theodor W. "The Curves of the Needle." Translated by Thomas Y. Levin. *October* 55 (1990): 49–55. https://doi.org/10.2307/778935.

Akrich, Madeleine. "The De-Scription of Technical Objects." In *Shaping Technology/Building Society: Studies in Sociotechnical Change*, edited by Wiebe E. Bijker and John Law, 205–24. Cambridge, MA: MIT Press, 1992.

Anand, Shaina. "10 Thesis [*sic*] on the Archive." In *Autonomous Archiving*, edited by Özge Çelikaslan, Alper Sen, and Pelin Tan, 79–94. Dpr-Barcelona, 2016.

Angulo, Jorge, Joan Calzada, and Alejandro Estruch. "Selection of Standards for Digital Television: The Battle for Latin America." *Telecommunications Policy* 35, no. 8 (September 2011): 773–87. https://doi.org/10.1016/j.telpol.2011.07.007.

Apple. "About Incompatible Media in Final Cut Pro X." Apple Support, October 7, 2019. https://support.apple.com/en-gb/HT209000.

Arthur, W. Brian. "Competing Technologies, Increasing Returns, and Lock-In by Historical Events." *The Economic Journal* 99, no. 394 (1989): 116–31. https://doi.org/10.2307/2234208.

Assmann, Aleida. "Canon and Archive." In *A Companion to Cultural Memory Studies*, edited by Astrid Erll and Ansgar Nünning, 97–108. Berlin/New York: De Gruyter, 2010.

Balbi, Gabriele, and Andreas Fickers, eds. *History of the International Telecommunication Union*. Berlin: De Gruyter, 2020.

Balsom, Erika. "From Singular to Plural." In *Accidental Archivism: Shaping Cinema's Futures with Remnants of the Past*, edited by Vinzenz Hediger and Stefanie Schulte Strathaus. Lüneburg: meson press, 2023.

Baron, Luis Fernando, and Ricardo Gomez. "The Associations between Technologies and Societies: The Utility of Actor-Network Theory." *Science, Technology and Society*, June 9, 2016. https://doi.org/10.1177/0971721816640615.

Bartlett, Albert A. "Algebra, the Golden Rectangle, and Paper Sizes." *Journal of College Science Teaching* 13, no. 6 (1984): 410–13.

Belton, J. "Historical Paper: The Origins of 35mm Film as a Standard." *SMPTE Journal* 99, no. 8 (August 1990): 652–61. https://doi.org/10.5594/J02613.

Berg, Marc. *Rationalizing Medical Work: Decision Support Techniques and Medical Practices*. Cambridge, MA: MIT Press, 1997.

Bischofberger, Catherine. "Moving Ahead with Gender Diversity at the IEC." *Standardization—Journal of Research and Innovation* 1, no. 3 (September 9, 2022): 1–3. https://doi.org/10.58831/2831-7920-1.3.1.

Bishop, Ann Peterson, and Susan Leigh Star. "Social Informatics of Digital Library Use and Infrastructure." *Annual Review of Information Science and Technology (ARIST)* 31 (1996): 301–401.

Blanchette, Jean-François. "A Material History of Bits." *Journal of the American Society for Information Science and Technology* 62, no. 6 (June 1, 2011): 1042–57. https://doi.org/10.1002/asi.21542.

Blanchette, Jean-François, and Deborah G. Johnson. "Data Retention and the Panoptic Society: The Social Benefits of Forgetfulness." *The Information Society* 18, no. 1 (January 1, 2002): 33–45. https://doi.org/10.1080/01972240252818216.

Blewer, Ashley, and Dave Rice. "Status of CELLAR: Update from an IETF Working Group for Matroska and FFV1." In *13th International Conference on Digital Preservation Proceedings*, 27–31. Bern: Swiss National Library, 2016. https://phaidra.univie.ac.at/detail/o:503184.

Blind, Knut, and Stephan Gauch. "Trends in ICT Standards: The Relationship between European Standardisation Bodies and Standards Consortia." *Telecommunications Policy* 32, no. 7 (August 1, 2008): 503–13. https://doi.org/10.1016/j.telpol.2008.05.004.

Blizzard, Christopher. "Firefox, YouTube and WebM—Mozilla Hacks." *Mozilla Hacks—the Web Developer Blog* (blog). Accessed August 25, 2023. https://hacks.mozilla.org/2010/05/firefox-youtube-and-webm.

Blood, George. "Refining Conversion Contract Specifications: Determining Suitable Digital Video Formats for Medium-Term Storage," 2011. https://www.digitizationguidelines.gov/audio-visual/documents/IntrmMastVidFormatRecs_20111001.pdf.

Blot-Wellens, Camille, and Caroline Fourier. "Minimum Viable Archiving." Conference paper presented at the 7th Eye International Conference: Global Audiovisual Archiving—Exchange of Knowledge and Practices, Amsterdam, 2022.

Blue Ribbon Task Force On Sustainable Digital Preservation And Access. "Sustainable Economics for a Digital Planet: Ensuring Long-Term Access to Digital Information," May 21, 2014. https://www.oclc.org/research/publications/2010/long-term-access-digital-information.html.

Bourdieu, Pierre, and Loïc J. D. Wacquant. *An Invitation to Reflexive Sociology*. Chicago: University of Chicago Press, 1992.

Bowker, Geoffrey C. *Memory Practices in the Sciences*. Cambridge, MA: MIT Press, 2008.

Bowker, Geoffrey C., and Susan Leigh Star. "Building Information Infrastructures for Social Worlds—The Role of Classifications and Standards." In *Community Computing and Support Systems: Social Interaction in Networked Communities*, edited by Toru Ishida, 231–48. Berlin: Springer, 1998. http://dl.acm.org/citation.cfm?id=646698.701381.

Bowker, Geoffrey C., and Susan Leigh Star. *Sorting Things Out: Classification and Its Consequences*. Cambridge, MA: MIT Press, 2000.

Bradley, Kevin. "Defining Digital Sustainability." *Library Trends* 56, no. 1 (2007): 148–63. https://doi.org/10.1353/lib.2007.0044.

Bradner, Scott O. "The Internet Standards Process—Revision 3." Request for Comments. Internet Engineering Task Force, October 1996. https://doi.org/10.17487/RFC2026.

Brooke, Sian. "'Condescending, Rude, Assholes': Framing Gender and Hostility on Stack Overflow." In *Proceedings of the Third Workshop on Abusive Language Online*, edited by Sarah T. Roberts, Joel Tetreault, Vinodkumar Prabhakaran, and Zeerak Waseem, 172—80. Florence, Italy: Association for Computational Linguistics, 2019. https://doi.org/10.18653/v1/W19-3519.

Brookey, Robert Alan. "The Format Wars: Drawing the Battle Lines for the Next DVD." *Convergence* 13, no. 2 (May 1, 2007): 199–211. https://doi.org/10.1177/1354856507075245.

Brunsson, Nils, and Bengt Jacobsson. "Following Standards." In *A World of Standards*, edited by Nils Brunsson and Bengt Jacobsson, 127–37. Oxford: Oxford University Press, 2000.

Brunswicker, Sabine, Jorge Andres Rodriguez, and Jonathan D. Wareham. "Governing Standardization through Formal and Informal Standard Development Organizations." *Academy of Management Proceedings* 2015, no. 1 (January 2015): 18596. https://doi.org/10.5465/ambpp.2015.18596abstract.

Bubestinger, Peter, Hermann Lewetz, and Marion Jaks. "Comparing Video Codecs and Containers for Archives," August 13, 2015. http://download.das-werkstatt.com/pb/mthk/info/video/comparison_video_codecs_container.html.

Bubestinger-Steindl, Peter. "File Formats for Audiovisual Preservation: How to Choose?" *CDA 2016 Formátové Výzvy LTP*, October 2016, 58–81.

Bubestinger-Steindl, Peter. "Review and Comparison of FFV1 versus Other Lossless Video Codecs for Long-Term Preservation." In *Sustainable Audiovisual Collections Through Collaboration*, edited by Rachael Stoeltje, Vicki Shively, George Boston, Lars Gaustad, and Dietrich Schüller, 153–59. Proceedings of the 2016 Joint Technical Symposium. Bloomington: Indiana University Press, 2017. https://doi.org/10.2307/j.ctt2005wh5.25.

Busch, Lawrence. *Standards: Recipes for Reality*. Cambridge, MA: MIT Press, 2011.

Bystydzienski, Jill M., and Sharon R. Bird, eds. *Removing Barriers: Women in*

Academic Science, Technology, Engineering, and Mathematics. Illustrated edition. Bloomington: Indiana University Press, 2006.

Caballar, Rina Diane. "Battle of the Video Codecs: Coding-Efficient VVC vs. Royalty-Free AV1–IEEE Spectrum." IEEE Spectrum: Technology, Engineering, and Science News, February 5, 2020. https://spectrum.ieee.org/battle-video-codecs-hevc-coding-efficiency-vvc-royalty-free-av1.

Calvert, Melodie, and Jennifer Terry, eds. *Processed Lives: Gender and Technology in Everyday Life*. London: Routledge, 1997. https://doi.org/10.4324/9780203990988.

Campbell, Savannah. "The Great Video Codec Debate: JPEG 2000 and FFV1." New York University, 2016. www.nyu.edu/tisch/preservation/program/student_work/2016fall/16f_1807_Campbell_a2.pdf.

Campbell, Timothy. *Wireless Writing in the Age of Marconi*. Minneapolis: University of Minnesota Press, 2006.

Capell, Laura. "Digitization as a Preservation Method for Damaged Acetate Negatives: A Case Study." *The American Archivist* 73, no. 1 (May 2010): 235–49. https://doi.org/10.17723/aarc.73.1.x381802g137421h3.

Cappa, Carolina. "Sur, paredón y después. La conferencia internacional Eye 2022." *Journal of Film Preservation*, no. 107 (October 2022): 29–35.

Cargill, Carl. *The Social Construction of Technological Systems : New Directions in the Sociology and History of Technology /*. Cambridge, MA: MIT Press, c1987.

Carse, Ashley. "Nature as Infrastructure: Making and Managing the Panama Canal Watershed." *Social Studies of Science* 42, no. 4 (2012): 539–63.

Casey, Mike. "Encoding and Wrapper Decisions and Implementation for Video Preservation Master Files." Indiana University, March 2017. mdpi.iu.edu/doc/MDPIwhitepaper.pdf.

Casey, Mike. "Why Media Preservation Can't Wait: The Gathering Storm," January 2015. https://scholarworks.iu.edu/dspace/handle/2022/27255.

Cath, Corinne. "Loud Men Talking Loudly: Exclusionary Cultures of Internet Governance." Amsterdam: Critical Infrastructure Lab, 2023.

CELLAR. "Codec Encoding for LossLess Archiving and Realtime Transmission," 2018. datatracker.ietf.org/wg/cellar/about/.

Cherchi Usai, Paolo. "The Demise of Digital (Print #1)." *Film Quarterly* 59, no. 3 (2006).

Chun, Wendy Hui Kyong. *Control and Freedom: Power and Paranoia in the Age of Fiber Optics*. Boca Raton: MIT Press, 2008.

Chun, Wendy Hui Kyong. "The Enduring Ephemeral, or the Future Is a Memory." *Critical Inquiry* 35, no. 1 (Autumn 2008): 148–71.

Cockburn, Cynthia, and Susan Ormrod. *Gender and Technology in the Making*. London: SAGE, 1993.

De Kosnik, Abigail. *Rogue Archives: Digital Cultural Memory and Media Fandom*. Cambridge, MA: MIT Press, 2016.

De Kosnik, Abigail. "The Collector Is the Pirate." *International Journal of Communication* 6 (April 12, 2012): 523–41.

Decherney, Peter. *Hollywood's Copyright Wars*. Reprint. New York: University Press Group Ltd, 2013.

Denegri-Knott, Janice. "Sinking the Online 'Music Pirates:' Foucault, Power and Deviance on the Web." *Journal of Computer-Mediated Communication* 9, no. 4 (2004). https://onlinelibrary.wiley.com/doi/full/10.1111/j.1083-6101.2004.tb00293.x.

Devlin, Bruce. "MXF—the Material eXchange Format." *EBU Technical Review*, 2002. https://tech.ebu.ch/docs/techreview/trev_291-devlin.pdf.

Devlin, Bruce, and Jim Wilkinson. *The MXF Book: An Introduction to the Material eXchange Format*. Burlington, MA: Focal Press, 2006. tech.ebu.ch/docs/techreview/trev_291-devlin.pdf.

Dijck, José van. *The Culture of Connectivity: A Critical History of Social Media*. Oxford University Press, 2013. https://doi.org/10.1093/acprof:oso/9780199970773.001.0001.

Doane, Mary Ann. "The Indexical and the Concept of Medium Specificity." *Differences* 18, no. 1 (May 1, 2007): 128–52. https://doi.org/10.1215/10407391-2006-025.

Donaldson, Devan Ray, and Elizabeth Yakel. "Secondary Adoption of Technology Standards: The Case of PREMIS." *Archival Science* 13, no. 1 (March 2013): 55–83. https://doi.org/10.1007/s10502-012-9179-0.

Drucker, Johanna. "Performative Materiality and Theoretical Approaches to Interface." *DHQ: Digital Humanities Quarterly*: 7, no. 1 (2013). http://www.digitalhumanities.org/dhq/vol/7/1/000143/000143.html.

Dunn, Arthur D. *Notes on the Standardization of Paper Sizes*. London: London Free Press Printing, 1969.

Edmondson, Ray. *Audiovisual Archiving: Philosophy and Principles*. 3rd ed. Paris: UNESCO, 2016. http://unesdoc.unesco.org/images/0024/002439/243973E.pdf.

Emerson, Lori. "How to Read a Network, or the Internet as Unfinished Demo." In *The Digital Reading Condition*, edited by Maria Engberg, Iben Have, and Birgitte Stougaard Pedersen, 98–107. New York: Routledge, 2022.

Ensmenger, Nathan. "'Beards, Sandals, and Other Signs of Rugged Individualism': Masculine Culture within the Computing Professions." *Osiris* 30, no. 1 (January 2015): 38—65. https://doi.org/10.1086/682955.

Ermoshina, Ksenia, and Francesca Musiani. "'Standardising by Running Code': The Signal Protocol and de Facto Standardisation in End-to-End Encrypted Messaging." *Internet Histories* 3, no. 3–4 (October 2, 2019): 343–63. https://doi.org/10.1080/24701475.2019.1654697.

FADGI (The Federal Agencies Digitization Guidelines Initiative) Audio-Visual Working Group. "Digital File Formats for Videotape Reformatting: Part 4. Detailed Matrix for Encodings (Multi-Page)." FADGI, December

2, 2014. http://www.digitizationguidelines.gov/guidelines/FADGI_VideoReFormatCompare_pt4_20141202.pdf.

Fahle, Oliver, Marek Jancovic, Elisa Linseisen, and Alexandra Schneider. "Medium | Format." *Zeitschrift Für Medienwissenschaft* 12, no. 1 (Issue 21) (2020): 10–19.

Fairbairn, Natasha, Maria Assunta Pimpinelli, and Thelma Ross. *The FIAF Moving Image Cataloguing Manual.* Edited by Linda Tadic. International Federation of Film Archives, 2016. https://doi.org/10.2307/j.ctt2005t2g.

Fallon, Tessa, Dave Rice, and Ashley Blewer. "Conch—Appendix on Standardization Exercises." MediaArea, March 3, 2015. https://github.com/MediaArea/MediaConch/releases/download/2015.03.14/MediaAreaConch_Appendix_Standardization.pdf.

Federal Agencies Digital Guidelines Initiative (FADGI). "AS-07: MXF Archive and Preservation Format Application Specification," September 2017. www.digitizationguidelines.gov/guidelines/AS-07_20170908.pdf.

Federal Agencies Digitization Guidelines Initiative (FADGI). "Audio-Visual Format Documentation Project: Background Paper Federal Agencies Audio-Visual Digitization Guidelines Working Group," October 2010. www.digitizationguidelines.gov/guidelines/FADGI-AV_AppSpecProj_Bkgd_101007.pdf.

Ferreira, Pedro. "MXF—a Progress Report." *EBU Technical Review*, no. 2010 Q3 (2010): 1–10.

FIAF Technical Commission. "Digitisation for Film Archives—Assorted Complications," 2012. https://www.fiafnet.org/images/tinyUpload/E-Resources/Commission-And-PIP-Resources/TC_resources/Digital%20Complications%20v1.1.pdf.

FIAT/IFTA. "Call for Presentations FIAT/IFTA World Conference Vienna 7–10 October 2015." FIAT/IFTA, May 13, 2015.

Fickers, Andreas. *"Politique de la grandeur" versus "Made in Germany": Politische Kulturgeschichte der Technik am Beispiel der PAL-SECAM-Kontroverse.* München: De Gruyter Oldenbourg, 2007.

Fleischhauer, Carl. "Comparing Formats for Video Digitization." Webpage. Library of Congress Blogs: The Signal—Digital Preservation, December 3, 2014. http://blogs.loc.gov/digitalpreservation/2014/12/comparing-formats-for-video-digitization/.

Fleischhauer, Carl. "Looking at Preservation from the Digital Library Perspective." *The Moving Image* 3, no. 2 (2003): 96–100. https://doi.org/10.1353/mov.2003.0027.

Fleischhauer, Carl. "User Needs and MXF Options: Preservation Planning and the AS-07 Specification," September 10, 2015. www.digitizationguidelines.gov/guidelines/AS-07_userNeeds_MXFoptions_20150910.pdf.

Fleischhauer, Carl, and Kevin Bradley. "IASA-TC 06 Guidelines for the Preservation of Video Recordings." International Association of Sound and Audiovisual Archives, 2018. www.iasa-web.org/tc06/guidelines-preservation-video-recordings.

Flower, Harriet I. *The Art of Forgetting: Disgrace and Oblivion in Roman Political Culture*. Chapel Hill, NC: University of North Carolina Press, 2011.

Fossati, Giovanna. "For a Global Approach to Audiovisual Heritage: A Plea for North/South Exchange in Research and Practice." *NECSUS: European Journal of Media Studies* 10, no. 2 (December 13, 2021): 127–33. https://doi.org/10.25969/MEDIAREP/17281.

Fossati, Giovanna. *From Grain to Pixel: The Archival Life of Film in Transition*. Amsterdam: Amsterdam University Press, 2009.

Foucault, Michel. "Politics and the Study of Discourse." In *The Foucault Effect: Studies in Governmentality*, edited by Graham Burchell, Colin Gordon, and Peter M. Miller, 53–73. London: Harvester Wheatsheaf, 1991.

Foucault, Michel. *The Archaeology of Knowledge*. Translated by A.M. Sheridan Smith. New York, NY: Pantheon Books, 1982.

Fox, Mary Frank, Deborah G. Johnson, and Sue V. Rosser, eds. *Women, Gender, and Technology*. First Edition. Urbana: University of Illinois Press, 2006.

Free Software and Open Formats: Virtual Immortality and Independence for Digital Archives. Serving the Cultural Heritage: Open Source Preservation Workshop. National Library of Sweden, Stockholm, 2016. https://www.youtube.com/watch?v=i9cLq2VWWLY.

Fresa, Antonella, Börje Justrell, and Claudio Prandoni. "Digital Curation and Quality Standards for Memory Institutions: PREFORMA Research Project." *Archival Science* 15, no. 2 (March 25, 2015): 191–216. https://doi.org/10.1007/s10502-015-9242-8.

Frick, Caroline. *Saving Cinema: The Politics of Preservation*. Oxford, New York: Oxford University Press, 2011.

Fuller, Matthew. *Media Ecologies: Materialist Energies in Art and Technoculture*. Cambridge, MA: MIT Press, 2005.

Galloway, Alexander R. *Protocol: How Control Exists after Decentralization*. Cambridge, MA: MIT Press, 2004.

Galloway, Alexander R., and Jason R. LaRivière. "Compression in Philosophy." *Boundary 2* 44, no. 1 (February 1, 2017): 125–47. https://doi.org/10.1215/01903659-3725905.

Gasser, Les. "The Integration of Computing and Routine Work." *ACM Transactions on Information Systems* 4, no. 3 (July 1, 1986): 205–25. https://doi.org/10.1145/214427.214429.

Geimer, Peter. "Das Bild als Spur. Mutmaßung über ein untotes Paradigma." In *Spur: Spurenlesen als Orientierungstechnik und Wissenskunst*, edited by Sybille Krämer, Werner Kogge, and Gernot Grube, Originalausgabe., 95–120. Frankfurt am Main: Suhrkamp Verlag, 2007.

Genus, Audley, and Anne-Marie Coles. "Rethinking the Multi-Level

Perspective of Technological Transitions." *Research Policy* 37, no. 9 (October 2008): 1436–45.

Gießmann, Sebastian, and Gabriele Schabacher. "Umwege und Umnutzung oder: Was bewirkt ein »Workaround«?" In *Umnutzung: alte Sachen, neue Zwecke,* edited by Stephan Habscheid, Gero Hoch, Hilde Schröteler-von Brandt, and Volker Stein, 13–26. Diagonal. Göttingen: V&R unipress, 2014.

Gilmour, Ian, and R. Justin Davila. "Lossless Video Compression for Archives: Motion JPEG2k and Other Options," 2005. pdfs.semanticscholar.org/a5c9/ec0c69767fb4803494477cb8e6d6249efcdc.pdf?_ga=2.140968754.1426998070.1537559612-405665357.1537559612.

Gitelman, Lisa. *"Raw Data" Is an Oxymoron.* MIT Press, 2013.

Gooday, Graeme, and James Sumner, eds. *By Whose Standards? Standardization, Stability and Uniformity in the History of Information and Electrical Technologies.* Vol. 28. History of Technology. London: Continuum, 2008.

Gracy, Karen F. "Moving Image Preservation and Cultural Capital." *Library Trends* 56, no. 1 (2007): 183–97.

Graham, Peter S. *Intellectual Preservation: Electronic Preservation of the Third Kind.* Washington, DC: Commission on Preservation and Access, 1994.

Hanseth, Ole, Edoardo Jacucci, Miria Grisot, and Margunn Aanestad. "Reflexive Standardization: Side Effects and Complexity in Standard Making." *MIS Quarterly* 30 (2006): 563–81. https://doi.org/10.2307/25148773.

Harcourt, Alison, George Christou, and Seamus Simpson. *Global Standard Setting in Internet Governance.* Oxford: Oxford University Press, 2020.

Hayles, N. Katherine. *How We Became Posthuman: Virtual Bodies in Cybernetics, Literature, and Informatics.* University of Chicago Press, 1999.

Hediger, Vinzenz. "The Original Is Always Lost: Film History, Copyright Industries and the Problem of Reconstruction." In *Cinephilia: Movies, Love and Memory,* edited by Marijke de Valck and Malte Hagener, 135–49. Amsterdam: Amsterdam University Press, 2005.

Heller, Franziska. "Warum Filmgeschichte? Wie die Digitalisierung unser Bild der Vergangenheit verändert." *Cinema,* no. 61 (2016): 12–24. https://doi.org/info:doi/10.5167/uzh-131627.

Heller, Franziska, and Barbara Flueckiger. "Zur Wertigkeit von Filmen. Retrodigitalisierung und Filmwissenschaft." *Montage AV* 19, no. 2 (2010): 138–52.

Heslop, Helen, Simon Davis, and Andrew Wilson. "An Approach to the Preservation of Digital Records." National Archives of Australia, December 2002.

Heß, Philipp. "SDG 5 and the Gender Gap in Standardization: Empirical Evidence From Germany." *Sustainability* 12, no. 20 (January 2020): 8699. https://doi.org/10.3390/su12208699.

Heydegger, Volker. "Analysing the Impact of File Formats on Data Integrity." In *Archiving 2008. Final Program and Proceedings,* 50–56. Bern: Society for Imaging Science and Technology, 2008. http://www.ingentaconnect.com/content/ist/ac/2008/00002008/00000001/art00011.

Holdsworth, Amy. "'Television Resurrections': Television and Memory." *Cinema Journal* 47, no. 3 (2008): 137–44.

Holstein, James A., and Jaber F. Gubrium. *The Active Interview*. SAGE Publications, Inc., 1995. https://doi.org/10.4135/9781412986120.

Holt, Niles R. "Wilhelm Ostwald's 'The Bridge.'" *The British Journal for the History of Science* 10, no. 2 (1977): 146–50.

Huffman, D. A. "Notes on Information-Lossless Finite-State Automata." *Il Nuovo Cimento (1955–1965)* 13, no. 2 (September 1, 1959): 397–405. https://doi.org/10.1007/BF02724674.

Hughes, Thomas P. "The Evolution of Large Technological Systems." In *The Social Construction of Technological Systems: New Directions in the Sociology and History of Technology*, edited by Wiebe E. Bijker, Trevor J. Pinch, and Thomas P. Hughes, 51–82. Cambridge, MA: MIT Press, 1987.

Hunter, Philip. "DVB Selects AV1 With AVS3 And VVC As Candidate Codecs For Standardization." The Broadcast Bridge, July 16, 2021. https://www.thebroadcastbridge.com/content/entry/17180/dvb-selects-av1-with-avs3-and-vvc-as-candidate-codecs-for-standardization.

IASA Technical Committee. *Guidelines for the Preservation of Video Recordings (TC 06)*. Edited by Carl Fleischhauer and Kevin Bradley. Revised version. International Association of Sound and Audiovisual Archives, 2019. https://www.iasa-web.org/sites/default/files/publications/IASA-TC_06-A_v2019.pdf.

IASA Technical Committee. *Handling and Storage of Audio and Video Carriers (TC 05)*. Edited by Dietrich Schüller and Albrecht Häfner. International Association of Sound and Audiovisual Archives, 2014. www.iasa-web.org/tc05/handling-storage-audio-video-carriers.

Jakobs, Kai. *Standardisation Processes in IT: Impact, Problems and Benefits of User Participation*. Vieweg Professional Computing. Braunschweig: Vieweg, 2000.

Jancovic, Marek. *A Media Epigraphy of Video Compression: Reading Traces of Decay*. Cham: Palgrave Macmillan, 2023.

Jancovic, Marek. "Ghosts of the Past: Frame Rates, Cranking and Access to Early Cinema." In *Exposing the Film Apparatus: The Film Archive as a Research Laboratory*, edited by Giovanna Fossati and Annie van den Oever, 75–82. Framing Film. Amsterdam: Amsterdam University Press, 2016.

Jancovic, Marek. "Lossless Compression and the Future of Memory." *Interactions: Studies in Communication & Culture* 8, no. 1 (2017): 45–61. https://doi.org/10.1386/iscc.8.1.45_1.

Jarczyk, Agathe, Reto Kromer, and David Pfluger. "Memoriav Recommendations Digital Archiving of Film and Video: Principles and Guidance." Edited by Yves Niederhäuser. Memoriav Association for the preservation of the audiovisual heritage of Switzerland, 2017. https://memoriav.ch/wp-content/uploads/2017/09/Empfehlungen_Digitale-Archivierung_Version1.1_EN.pdf.

Jenkins, Henry. *Convergence Culture: Where Old and New Media Collide*. New York, NY: NYU Press, 2006.

Jones, Jimi, and Landon Palmer. "A Record for Your Television: The Capacitance Electronic Disc's History of Failure." *The Moving Image* 18, no. 2 (2019): 103–31.

Jong, Annemieke de. "Sound and Vision—Digital Preservation: Policy, Standards and Procedures." Netherlands Institute for Sound and Vision, January 2016.

Kaghan, William N., and Geoffrey C. Bowker. "Out of Machine Age?: Complexity, Sociotechnical Systems and Actor Network Theory." *Journal of Engineering and Technology Management* 18, no. 3–4 (September 2001): 253–69. https://doi.org/10.1016/S0923-4748(01)00037-6.

Kaufman, Peter B. "Towards a New Audiovisual Think Tank for Audiovisual Archivists and Cultural Heritage Professionals," January 2018. publications.beeldengeluid.nl/pub/574/TowardsANewThinkTankForAudiovisualArchivistsAndCulturalHeritageProfessionals_SoundAndVision_Jan2018.pdf.

Keenan, Thomas. "Counter-Forensics and Photography." *Grey Room* 55 (April 1, 2014): 58–77. https://doi.org/10.1162/GREY_a_00141.

Keitel, Christian. "Der nestor-Leitfaden zur Digitalen Bestandserhaltung und seine Folgen für die Archive." In *Digitale Archivierung in der Praxis: 16. Tagung des Arbeitskreises "Archivierung von Unterlagen aus Digitalen Systemen" und Nestor-Workshop "Koordinierungsstellen,"* edited by Christian Keitel and Kai Naumann, 267–77. Werkhefte der Staatlichen Archivverwaltung Baden-Württemberg : Serie A, H. 24. Stuttgart: Kohlhammer, 2013.

Kennelly, Arthur Edwin. *Artificial Electric Lines: Their Theory, Mode of Construction and Uses*. New York: McGraw-Hill Book Company, 1917.

Kerntopf, Paweł, Radomir Stanković, Alexis De Vos, and Jaakko Astola. "Early Pioneers to Reversible Computation." In *Post-Binary ULSI Systems, 23rd International Workshop, Proceedings*, edited by Naoyuki Homma and Hirofumi Nakahara, 36–42. Bremen: Japan Research Group on Multiple-Valued Logic, 2014. http://hdl.handle.net/1854/LU-4400338.

Kinross, Robin. *A4 and before: Towards a Long History of Paper Sizes*. KB Lecture 6. Wassenaar: NIAS, 2009.

Kirschenbaum, Matthew. *Mechanisms: New Media and the Forensic Imagination*. Cambridge, MA: MIT Press, 2012.

Kittler, Friedrich A. "There Is No Software." CTheory.net, October 18, 1995. http://www.ctheory.net/articles.aspx?id=74.

Klein, Hans K., and Daniel Lee Kleinman. "The Social Construction of Technology: Structural Considerations." *Science, Technology, & Human Values* 27, no. 1 (January 1, 2002): 28–52. https://doi.org/10.1177/016224390202700102.

Klerk, Christo de. "Materiality of Deletion." *MAS CONTEXT*, Surveillance, no. 22 (Summer 2014): 183–93.

Kling, Rob. "Learning About Information Technologies and Social Change:

The Contribution of Social Informatics." *The Information Society* 16, no. 3 (2000): 217–32.

Kling, Rob. "What Is Social Informatics and Why Does It Matter?" *The Information Society* 23, no. 4 (June 21, 2007): 205–20. https://doi.org/10.1080/01972240701441556.

Kokaram, A.C. "On Missing Data Treatment for Degraded Video and Film Archives: A Survey and a New Bayesian Approach." *IEEE Transactions on Image Processing* 13, no. 3 (March 2004): 397–415. https://doi.org/10.1109/TIP.2004.823815.

Koretz, Daniel. *Measuring Up: What Educational Testing Really Tells Us*. 1st edition. Cambridge, MA: Harvard University Press, 2008.

Kozel, Susan, ed. *Openness: Politics, Practices, Poetics*. Malmö: Malmö University, The Living Archives Project, 2017.

Krajewski, Markus. *Restlosigkeit: Weltprojekte um 1900*. Frankfurt am Main: Fischer Taschenbuch, 2006.

Kromhout, Melle Jan. "Noise Resonance: Technological Sound Reproduction and the Logic of Filtering." Doctoral dissertation, University of Amsterdam, 2017.

Kuhr, Stefan. "Implementation of a JPEG Decoder for a 16-Bit Microcontroller." Master's thesis, University of Applied Sciences, 2001. https://www.stefan-kuhr.de/download/thesis2side.pdf.

Kulovits, Hannes, Andreas Rauber, Anna Kugler, Markus Brantl, Tobias Beinert, and Astrid Schoger. "From TIFF to JPEG 2000?: Preservation Planning at the Bavarian State Library Using a Collection of Digitized 16th Century Printings." *D-Lib Magazine* 15, no. 11/12 (November 2009). https://doi.org/10.1045/november2009-kulovits.

Kvale, Steinar. *InterViews: An Introduction to Qualitative Research Interviewing*. InterViews: An Introduction to Qualitative Research Interviewing. Thousand Oaks, CA, US: Sage Publications, Inc, 1994.

Kwak, Jooyoung, Heejin Lee, and Vladislav V. Fomin. "Government Coordination of Conflicting Interests in Standardisation: Case Studies of Indigenous ICT Standards in China and South Korea." *Technology Analysis & Strategic Management* 23, no. 7 (August 1, 2011): 789–806. https://doi.org/10.1080/09537325.2011.592285.

Larkin, Brian. *Signal and Noise: Media, Infrastructure, and Urban Culture in Nigeria*. Durham: Duke University Press Books, 2008.

Larkin, Brian. "The Materiality of Cinema Theaters in Northern Nigeria." In *Media Worlds: Anthropology on New Terrain*, edited by Faye D. Ginsburg, Lila Abu-Lughod, and Brian Larkin, 319–36. University of California Press, 2002.

Law, John. "Notes on the Theory of the Actor-Network: Ordering, Strategy, and Heterogeneity." *Systems Practice* 5, no. 4 (August 1, 1992): 379–93. https://doi.org/10.1007/BF01059830.

Lawrence, Gregory W., William R. Kehoe, Oya Y. Rieger, William H. Walters, and Anne R. Kenney. *Risk Management of Digital Information: A File Format Investigation*. Washington, D.C.: Council on Library and Information Resources, 2000.

Lesk, Michael. "Preservation of New Technology. A Report of the Technology Assessment Advisory Committee to the Commission on Preservation and Access." Commission on Preservation and Access, October 1992.

Lischer-Katz, Zack. "Considering JPEG2000 for Video Preservation: A Battle for Epistemic Ground." In *iConference 2014 Proceedings*, 1056–65. iSchools, 2014. https://doi.org/10.9776/14381.

Lischer-Katz, Zack. "Studying the Materiality of Media Archives in the Age of Digitization: Forensics, Infrastructures and Ecologies." *First Monday* 22, no. 1 (2017). https://doi.org/10.5210/fm.v22i1.7263.

Lobato, Ramon. *Shadow Economies of Cinema: Mapping Informal Film Distribution*. London: British Film Institute, 2012.

Lomonosov Moscow State University Graphics & Media Lab (Video Group). "Lossless Video Codecs Comparison '2007." Compression.ru, 2007. http://compression.ru/video/codec_comparison/lossless_codecs_2007_en.html.

Lorrain, Emanuel. "A Short Guide to Choosing a Digital Format for Video Archiving Masters," March 2014. www.scart.be/?q=en/content/short-guide-choosing-digital-format-video-archiving-masters.

Lorrain, Emanuel. "Interview with Peter Bubestinger and Herman Lewetz (Österreichische Mediathek)," May 6, 2014. https://www.scart.be/?q=en/content/interview-peter-bubestinger-and-herman-lewetz-%C3%B6sterreichische-mediathek.

Lovink, Geert, and Ned Rossiter. "Seriality for All: The Role of Protocols and Standards in Critical Theory." In *Interoperabel Nederland*, edited by Peter Waters, Nico Westpalm van Hoorn, and Pieter Wisse, 426–34. The Hague: Forum Standardisatie, 2011.

Mackenzie, Adrian. "Every Thing Thinks: Sub-Representative Differences in Digital Video Codecs." In *Deleuzian Intersections: Science, Technology, Anthropology*, edited by Casper Bruun Jensen and Kjetil Rodje, 139–54. New York/Oxford: Berghahn Books, 2013.

Mackenzie, Adrian. *Wirelessness: Radical Empiricism in Network Cultures*. Cambridge, MA: MIT Press, 2010.

MacKenzie, Donald, and Judy Wajcman, eds. *The Social Shaping of Technology*. 1999th ed. Buckingham [England]: Open University Press. Accessed August 28, 2023. https://www.amazon.com/Social-Shaping-Technology-Donald-MacKenzie/dp/0335199135.

Manovich, Lev. *The Language of New Media*. Cambridge, MA: MIT Press, 2001.

Marsh, Alex. "The Pros and Cons of FFV1." *Duke Universities Library: Bitstreams—The Digital Collections Blog* (blog), May 8, 2015. http://blogs.library.duke.edu/bitstreams/2015/05/08/the-pros-and-cons-of-ffv1/.

Mason, Mark. "What Is Complexity Theory and What Are Its Implications for

Educational Change?" *Educational Philosophy and Theory* 40, no. 1 (2008): 35–49. https://doi.org/10.1111/j.1469-5812.2007.00413.x.

Mayer-Schönberger, Viktor. *Delete: The Virtue of Forgetting in the Digital Age*. Princeton: Princeton University Press, 2009.

McGill, Meredith L. "Format." *Early American Studies: An Interdisciplinary Journal* 16, no. 4 (2018): 671–77. https://doi.org/10.1353/eam.2018.0033.

Menkman, Rosa. "A Vernacular of File Formats—A Guide to Databend Compression Design," August 2010. https://beyondresolution.info/A-Vernacular-of-File-Formats.

Menkman, Rosa. "Behind the White Shadows of Image Processing: Shirley, Lena, Jennifer and the Angel of History." In *Faceless: Re-Inventing Privacy Through Subversive Media Strategies*, edited by Bogomir Doringer and Brigitte Felderer, 159–79. Berlin: Walter de Gruyter, 2018.

Meyer, E.T. *Examining the Hyphen: The Value of Social Informatics for Research and Teaching. In Social Informatics: Past, Present and Future*. Edited by Pnina Fichman and Howard Rosenbaum. Newcastle Upon Tyne, England: Cambridge Scholars Publishing, 2014.

Mosco, Vincent. *The Political Economy of Communication: Rethinking and Renewal*. Thousand Oaks, CA: Sage Publications, 1996.

Mulvin, Dylan. *Proxies: The Cultural Work of Standing In*. Cambridge, MA: MIT Press, 2021.

Mulvin, Dylan, and Jonathan Sterne. "Scenes from an Imaginary Country: Test Images and the American Color Television Standard—Dylan Mulvin, Jonathan Sterne, 2016." *Television & New Media* 17, no. 1 (2016): 21–43.

Murphy, Craig N., and Joanne Yates. *The International Organization for Standardization (ISO): Global Governance Through Voluntary Consensus*. London: Routledge, 2008. https://doi.org/10.4324/9780203884348.

National Digital Stewardship Alliance. "National Agenda For Digital Stewardship," 2015. www.digitalpreservation.gov/documents/2015NationalAgenda.pdf.

Niedermayer, Michael. "FFV1 Video Codec Specification," 2013. http://ffmpeg.org/~michael/ffv1.html#huffman-coding-mode.

Njoroge, Njoroge, Rich Rath, and Ted Sammons. "The Work of Vinyl in the Age of Digital Reproduction." *American History Now—Explorations in Digital Curation* (blog), March 24, 2014. http://americanhistorynow.org/2014/03/24/work-of-vinyl-in-the-age-of-digital-reproduction/.

Nowak, A., and S. Fößel. "Proposal of a System Architecture for Digital Film Archives Using JPEG 2000 and MXF." *SMPTE Motion Imaging Journal* 117, no. 5 (2008): 42–48.

Obiaya, Ikechukwu. "A Break with the Past: The Nigerian Video-Film Industry in the Context of Colonial Filmmaking." *Film History* 23, no. 2 (2011): 129–46. https://doi.org/10.2979/filmhistory.23.2.129.

Oever, Niels ten, and Stefania Milan. "The Making of International Communication Standards: Towards a Theory of Power in Standardization." *Journal of Standardisation* 1 (July 6, 2022). https://doi.org/10.18757/jos.2022.6205.

Olayiwola, Abiodun. "From Celluloid to Video: The Tragedy of the Nigerian Film Industry." *Journal of Film and Video* 59, no. 3 (2007): 58–61.

O'Leary, Kieran. "Introduction to FFV1 and Matroska for Film Scans." *Digital AV Preservation Scripts* (blog), October 7, 2016. https://kieranjol.wordpress.com/2016/10/07/introduction-to-ffv1-and-matroska-for-film-scans/.

Opoku-Boateng, Judith, Ekow Cann, Samuel Aniegye Ntewusu, and Sandra Owusu. "The J.H. Kwabena Nketia Archive at the University of Ghana-Legon." *History in Africa* 47 (June 2020): 375–82. https://doi.org/10.1017/hia.2019.27.

Parkouda, Michelle, and Diane (Xiaolu) Liao. "Understanding the Relationship Between the Gender Composition of Technical Committees and Performance." *Standardization—Journal of Research and Innovation* 3, no. 2 (March 5, 2024): 1–5. https://doi.org/10.58831/2831-7920-3.2.1.

Pavuza, Franz. "Migration of Archived Video Footage: Challenges and Solutions for Hardware, Software, and Workflow Issues." In *Sustainable Audiovisual Collections Through Collaboration*, edited by Rachael Stoeltje, Vicki Shively, George Boston, Lars Gaustad, and Dietrich Schüller, 60–63. Proceedings of the 2016 Joint Technical Symposium. Bloomington: Indiana University Press, 2017. https://doi.org/10.2307/j.ctt2005wh5.13.

Pedersen, Mogens Kühn, Vladislav V. Fomin, and Henk J. de Vries. "Open Standards and Government Policy." In *Information Communication Technology Standardization for E-Business Sectors: Integrating Supply and Demand Factors*, edited by Kai Jakobs, 188–99. Hershey, PA: IGI Global, 2009. https://doi.org/10.4018/978-1-60566-320-3.ch013.

Petty, Sheila. "Digital Video Films as 'Independent' African Cinema." In *Independent Filmmaking Around the Globe*, edited by Doris Baltruschat and Mary P. Erickson, 255–69. University of Toronto Press, 2015. https://www.jstor.org/stable/10.3138/j.ctvg254pb.18.

Philpott, Eric. "Changes to MKV Format Support in Premiere Pro and Adobe Media Encoder." Adobe Blog, March 5, 2019. https://blog.adobe.com/en/publish/2019/05/03/changes-to-mkv-format-support-in-premiere-pro-and-adobe-media-encoder.

Pickren, Graham. "Political Ecologies of Electronic Waste: Uncertainty and Legitimacy in the Governance of E-Waste Geographies." *Environment and Planning A: Economy and Space* 46, no. 1 (January 1, 2014): 26–45. https://doi.org/10.1068/a45728.

Pinch, Trevor. "Giving Birth to New Users: How the Mini-Moog Was Sold to Rock and Roll." In *How Users Matter: The Co-Construction of Users and Technology*, edited by Nelly Oudshoorn and Trevor Pinch, 247–70. Cambridge, MA: The MIT Press, 2005.

Pinch, Trevor, and Wiebe E. Bijker. "The Social Construction of Facts and Artifacts: Or How the Sociology of Science and the Sociology of Technology Might Benefit Each Other." In *The Social Construction of Technological*

Systems: New Directions in the Sociology and History of Technology, edited by Wiebe E. Bijker, Thomas P. Hughes, and Trevor Pinch, 17–50. Cambridge, MA: MIT Press, 1987.

Prell, Christina. "Rethinking the Social Construction of Technology through 'Following the Actors': A Reappraisal of Technological Frames." *Sociological Research Online* 14, no. 2 (March 1, 2009): 36–47. https://doi.org/10.5153/sro.1913.

Prentice, Will, and Lars Gaustad, eds. *The Safeguarding of the Audiovisual Heritage: Ethics, Principles and Preservation Strategy (IASA-TC 03)*. 4th ed. London: International Association of Sound Archives and Audiovisual Archives, 2017.

Quinn, Terry. *From Artefacts to Atoms: The BIPM and the Search for Ultimate Measurement Standards*. Oxford: Oxford University Press, 2011.

Rakemane, Donald, and Olefhile Mosweu. "Challenges of Managing and Preserving Audio-Visual Archives in Archival Institutions in Sub Saharan Africa: A Literature Review." *Collection and Curation* 40, no. 2 (January 1, 2020): 42–50. https://doi.org/10.1108/CC-04-2020-0011.

Reagle, Joseph. "'Free as in Sexist?' Free Culture and the Gender Gap." *First Monday* 18, no. 1 (2012). https://doi.org/10.5210/fm.v18i1.4291.

Rice, Dave. "Reconsidering the Checksum for Audiovisual Preservation: Detecting Digital Change in Audiovisual Data with Decoders and Checksums." *IASA Journal*, no. 39 (June 2012): 34–40.

Robinson, Helena. "Remembering Things Differently: Museums, Libraries and Archives as Memory Institutions and the Implications for Convergence." *Museum Management and Curatorship* 27, no. 4 (October 2012): 413–29. https://doi.org/10.1080/09647775.2012.720188.

Rossell, Deac. "Die soziale Konstruktion früher technischer Systeme der Filmprojektion." *KINtop*, no. 8 (1999): 53–81.

Roth, Lorna. "Looking at Shirley, the Ultimate Norm: Colour Balance, Image Technologies, and Cognitive Equity." *Canadian Journal of Communication* 34, no. 1 (March 28, 2009): 111–36.

Rubin, N. "Preserving Digital Public Television Final Report." Library of Congress, June 2010. www.digitalpreservation.gov/partners/documents/pdpt_finalreport0610.pdf.

Saager, Adolf, and Karl Wilhelm Bührer. *Die Organisierung der geistigen Arbeit durch "die Brücke", von K.W. Bührer und Ad. Saager*. Ansbach: F. Seybold, 1911.

Schäfer, Heinrich Wilhelm. *HabitusAnalysis 2—Praxeology and Meaning*. Wiesbaden: Springer Nature, 2020.

Schivelbusch, Wolfgang. *The Railway Journey: The Industrialization of Time and Space in the Nineteenth Century*, 2014.

Schlaudt, Oliver, and Lara Huber, eds. *Standardization in Measurement: Philosophical, Historical and Sociological Issues*. 1st edition. London: Routledge, 2015.

Schmidt, Robert. "The Methodological Challenges of Practicising Praxeology."

In *Practice Theory and Research: Exploring the Dynamics of Social Life*, edited by Gert Spaargaren, Don Weenink, and Machiel Lamers, 43–59. Abingdon / New York: Routledge, 2016.

Schmidt, Susanne K., and Raymund Werle. *Coordinating Technology: Studies in the International Standardization of Telecommunications*. Cambridge, MA: The MIT Press, 1997.

Schmidt-Bachem, Heinz. *Aus Papier: eine Kultur- und Wirtschaftsgeschichte der Papier verarbeitenden Industrie in Deutschland*. Berlin/Boston: Walter de Gruyter, 2011.

Schneider, Alexandra. "Viewer's Digest: Small Gauge and Reduction Prints as Liminal Compression Formats." In *Format Matters Standards, Practices, and Politics in Media Cultures*, edited by Marek Jancovic, Axel Volmar, and Alexandra Schneider, 129–46. Lüneburg: meson press, 2020.

Schüttpelz, Erhard. "Die medientechnische Überlegenheit des Westens. Zur Geschichte und Geographie der immutable mobiles Bruno Latours." In *Mediengeographie: Theorie—Analyse—Diskussion*, edited by Jörg Döring and Tristan Thielmann, 67–110. Bielefeld: transcript, 2009.

The Science and Technology Council of the Academy of Motion Picture Arts and Sciences. "The Digital Dilemma: Strategic Issues in Archiving and Accessing Digital Motion Picture Materials." The Academy of Motion Picture Arts and Sciences, November 2007. www.oscars.org/science-technology/council/projects/digitaldilemma/.

Slide, Anthony. *Nitrate Won't Wait: A History of Film Preservation in the United States*. Jefferson, North Carolina: McFarland Press, 1992.

Smith, Abby. "Valuing Preservation." *Library Trends* 56, no. 1 (2007): 4–25.

Smith, Trudi Lynn, and Kate Hennessy. "Anarchival Materiality in Film Archives: Toward an Anthropology of the Multimodal." *Visual Anthropology Review* 36, no. 1 (March 2020): 113–36. https://doi.org/10.1111/var.12196.

Smither, Roger. "Formats and Standards: A Film Archive Perspective on Exchanging Computerized Data." *The American Archivist* 50, no. 3 (July 1987): 324–37. https://doi.org/10.17723/aarc.50.3.5802724670213420.

Smolicki, Jacek. "Para-Archives : Rethinking Personal Archiving Practices in the Times of Capture Culture." Doctoral dissertation, Malmö University, 2017.

Snyder, James. "JPEG2000 in Moving Image Archiving." Presented at the JPEG2000 Summit, National National Audio Visual Conservation Conservation Center Culpeper, Virginia, May 13, 2011. http://www.digitizationguidelines.gov/still-image/documents/Snyder.pdf.

Stanco, Filippo, Sebastiano Battiato, and Giovanni Gallo. *Digital Imaging for Cultural Heritage Preservation Analysis, Restoration, and Reconstruction of Ancient Artworks*. Boca Raton: CRC Press, 2017.

Star, Susan Leigh. "The Ethnography of Infrastructure." *American Behavioral Scientist* 43, no. 3 (November 1999): 377–91. https://doi.org/10.1177/00027649921955326.

Star, Susan Leigh. "The Structure of Ill-Structured Solutions: Boundary Objects and Heterogeneous Distributed Problem Solving." In *Distributed Artificial Intelligence*, 37–54. Elsevier, 1989. https://doi.org/10.1016/B978-1-55860-092-8.50006-X.

Star, Susan Leigh, and James R. Griesemer. "Institutional Ecology, 'Translations' and Boundary Objects." *Amateurs and Professionals in Social Studies of Science* 19, no. 0 (1989): 387–420.

Star, Susan Leigh, and Martha Lampland. "Reckoning with Standards." In *Standards and Their Stories: How Quantifying, Classifying, and Formalizing Practices Shape Everyday Life*, edited by Martha Lampland and Susan Leigh Star, 3–34. Ithaca, NY: Cornell University Press, 2008.

Stauff, Markus. "Das neue Fernsehen. Machteffekte einer heterogenen Kulturtechnologie." Dissertation, Ruhr-Universität, 2004.

Sterne, Jonathan. *MP3: The Meaning of a Format*. Durham, NC: Duke University Press, 2012.

Sterne, Jonathan. "Out with the Trash: On the Future of New Media." In *Residual Media*, edited by Charles R. Acland, 16–31. Minneapolis: University of Minnesota Press, 2007.

Suchman, Lucy, J. E. Blomberg, and R. Trigg. "Reconstructing Technologies as Social Practice." *American Behavioral Scientist* 43, no. 3 (1999): 392–408.

Sugimoto, Cassidy R., and Vincent Larivière. *Equity for Women in Science: Dismantling Systemic Barriers to Advancement*. Cambridge, MA: Harvard University Press, 2023.

Tansey, Eira. "Archival Adaptation to Climate Change." *Sustainability: Science, Practice and Policy* 11, no. 2 (November 1, 2015): 45–56. https://doi.org/10.1080/15487733.2015.11908146.

Taubman, David S., and Michael W. Marcellin. *JPEG 2000: Image Compression Fundamentals, Standards and Practice*. Norwell, MA: Kluwer Academic Publishers, 2002.

Thylstrup, Nanna Bonde. *The Politics Of Mass Digitization*. Cambridge, MA: The MIT Press, 2018.

Timmermans, Stefan, and Steven Epstein. "A World of Standards but Not a Standard World: Toward a Sociology of Standards and Standardization." *Annual Review of Sociology* 36 (2010): 69–89.

Turquety, Benoît. *Medium, Format, Configuration: The Displacements of Film*. Lüneburg: meson press, 2019. https://meson.press/books/medium-format-configuration/.

Van House, N., and E. F. Churchill. "Technologies of Memory: Key Issues and Critical Perspectives." *Memory Studies* 1, no. 3 (2008): 295–310.

Villarreal, Andrea Barrios. *International Standardization and the Agreement on Technical Barriers to Trade*. Cambridge: Cambridge University Press, 2018.

Volmar, Axel. "Das Format als medienindustriell motivierte Norm: Überlegungen zu einem medienkulturwissenschaftlichen Formatbegriff." *Zeitschrift für Medienwissenschaft* 12, no. 1 (2020): 19–30.

Volmar, Axel."Formats as Media of Cooperation." *Media in Action* 1, no. 2 (2017): 9–28.

Volmar, Axel, Marek Jancovic, and Alexandra Schneider. "Format Matters: An Introduction to Format Studies." In *Format Matters: Standards, Practices, and Politics in Media Cultures*, edited by Marek Jancovic, Axel Volmar, and Alexandra Schneider, 7–26. Lüneburg: meson press, 2020.

Vries, H. J. de. *Standardization—A Business Approach to the Role of National Standardization Organizations*. Boston, MA: Kluwer Academic Publishers, 2003. https://www.amazon.com/Standardization-Business-Approach-National-Organizations/dp/0792386388.

Wajcman, Judy. *Feminism Confronts Technology*. University Park: Penn State University Press, 1991.

Walters, Tyler. "Thinking About Archival Preservation in the '90s and Beyond: Some Recent Publications and Their Implications for Archivists." *The American Archivist* 58, no. 4 (September 1995): 476–93. https://doi.org/10.17723/aarc.58.4.t05414u2q067767g.

Ward, Alan. *A Manual of Sound Archive Administration*. Aldershot: Gower Pub. Co, 1990.

Wasser, Frederick. *The Hollywood Empire and the VCR*. Austin: University of Texas Press, 2001.

Wasson, Haidee. "Formatting Film Studies." *Film Studies* 12, no. 1 (May 1, 2015): 57–61. https://doi.org/10.7227/FS.12.0007.

Waters, Donald J., Anne R. Kenney, Lynne K. Personius, M. Stuart Lynn, and Millicent D. Abell. "The Digital Preservation Consortium Mission and Goals." Commission on Preservation and Access, 1994.

Weber, Anne-Katrin. "Recording on Film, Transmitting by Signals: The Intermediate Film System and Television's Hybridity in the Interwar Period." *Grey Room* 56 (July 2014): 6–33. https://doi.org/10.1162/GREY_a_00148.

Werle, Raymund. "Standards in the International Telecommunications Regime." Institut für Wirtschaftsforschung-Hamburg, Hamburgische Welt-Wirtschafts-Archiv, January 2001. ageconsearch.umn.edu/bitstream/26265/1/dp010157.pdf.

Whitehead, Anne. *Memory*. London; New York: Routledge, 2008.

Winner, Langdon. "Do Artifacts Have Politics?" *Daedelus* 109, no. 1 (1980): 121–36.

Woolgar, Steve. "Configuring the User: The Case of Usability Trials." *The Sociological Review* 38, no. S1 (1990): 58–99. https://doi.org/10.1111/j.1467-954X.1990.tb03349.x.

Wright, Richard. "Preserving Moving Pictures and Sound," 2012. www.dpconline.org/docman/technology-watch-reports/753-dpctw12-01-pdf/file.

Yates, JoAnne, and Craig N. Murphy. *Engineering Rules: Global Standard Setting since 1880*. Baltimore: Johns Hopkins University Press, 2019.

Yocum, Josh. "Audio/Video Digitization Guidelines for Northwestern Library." Unpublished, 2016.

Interviews

Akporherhe, Justina. Personal interview, March 22, 2021.

Berney, Tre. Personal interview, April 19, 2023.

Bird, Jim. Personal interview, March 22, 2018.

Blewer, Ashley. Personal interview, September 13, 2017; e-mail interview, April 6, 2023.

Bubestinger-Steindl, Peter. Personal interview, October 17, 2017.

Casey, Mike. Personal interview, March 23, 2018.

DaPuzzo, Andrew. Personal interview, March 23, 2018.

Declercq, Brecht. Personal interview, September 22, 2017.

Dix, Erik. Personal interview, September 8, 2017.

Dunn, John. Personal interview, March 23, 2018.

Gant, Anne. Personal interview, August 18, 2017.

Gaustad, Lars. Personal interview, December 20, 2017.

Henderson, Ian. Personal interview, September 21, 2017.

Jaks, Marion. Personal interview, November 3, 2017.

Joshua, Bright. Personal interview, May 21, 2021.

Krabbenhoeft, Nick. Personal interview, January 18, 2018.

Kromer, Reto. Personal interview, October 18, 2017.

Kummer, Jean-Christophe. Personal interview, March 23, 2018.

Lewetz, Hermann. Personal interview, December 1, 2017.

Lhomme, Steve. Personal interview, December 1, 2017.

McLellan, Evelyn. Personal interview, September 20, 2017.

Mobley, Rob. Personal interview, March 23, 2018.

Murray, Kate. Personal interview, September 28, 2017.

Noorlander, Ivo. Personal interview, August 18, 2017.

Opoku-Boateng, Judith. Personal interview, August 26, 2021.

Rice, Dave. Personal interviews, September 8, 2017, and February 21, 2018.

Schupert, Jim. Personal interview, March 22, 2018.

Snyder, James. Personal interview, April 27, 2018.

Turkus, Benjamin. Personal interview, September 29, 2017.

Verbruggen, Erwin. Personal interview, June 25, 2019.

Wheeler, Brian. Personal interview, March 23, 2018.

Yocum, Josh. Personal interview, October 4, 2017.

Index

JIMI JONES is an adjunct lecturer of library and information sciences for the School of Information Sciences at the University of Illinois Urbana-Champaign.

MAREK JANCOVIC is an assistant professor of media studies at Vrije Universiteit Amsterdam. He is the author of *A Media Epigraphy of Video Compression: Reading Traces of Decay*.

The University of Illinois Press
is a founding member of the
Association of University Presses.

Composed in 10.5/13 Mercury Text
with Avenir display
by Jim Proefrock
at the University of Illinois Press

University of Illinois Press
1325 South Oak Street
Champaign, IL 61820-6903
www.press.uillinois.edu